Postharvest Diseases of Perishables and its Management

THE EDITOR

Dr. K. Prabakar, Director (CPPS) , TNAU, Coimbatore, Completed his, M.Sc.(Ag) during 1986 and joined as Assistant Professor (Plant Pathology) in Tamil Nadu Agricultural University during 1988. He is specialized in postharvest and seed pathology. He has teaching experience of more than 30 years. He handled classes for UG, PG and Ph.D. students. He has guided M.Sc. (Ag) and Ph.D. students of Plant Pathology and at present guiding Ph.D. students. He has received TNAU Best researcher award and Best book award from MASU apart from other awards. He has obtained externally funded projects worth more than 10 crores. He was a FAO Consultant for Afghanistan for setting up plant health inspectorate. He has also instrumental in the release of one betelvine and one sugarcane variety.

Postharvest Diseases of Perishables and its Management

K. Prabakar

Director,
Centre for Plant Protection Studies,
Tamil Nadu Agricultural University,
Coimbatore – 3, Tamil Nadu

2022

Daya Publishing House®
A Division of
Astral International Pvt. Ltd.
New Delhi – 110 002

ISBN: 978-93-90435-92-0 (HB)

Published by : **Daya Publishing House®**
A Division of
Astral International Pvt. Ltd.
– ISO 9001:2015 Certified Company –
4736/23, Ansari Road, Darya Ganj,
New Delhi-110 002
Ph. 011-43549197, 23278134
E-mail: info@astralint.com
Website: www.astralint.com

Digitally Printed at : **Replika Press Pvt. Ltd.**

TAMIL NADU AGRICULTURAL UNIVERSITY

Prof. N. KUMAR, Ph.D., F.H.S.I., F.S.P.H.

Vice-Chancellor

Coimbatore - 641003
Tamil Nadu, India

FOREWORD

Agricultural and horticultural crop produces are severely affected by many post-harvest diseases during transit and storage. The achievements made in food production in developing countries through green revolution can be maintained only if proper importance is given for post harvest operations especially related to handling, storage and value addition. In recent years, new techniques have been developed for detection, diagnosis and management of post-harvest diseases and advanced storage methods for improving the value of stored products.

A well-documented comprehensive book entitled "Post-harvest diseases of perishables and its management" has been brought out by Dr. K.Prabakar, Director, Centre for Plant Protection Studies will help the students, stakeholders and consumers to save their crops from post-harvest losses through identification of the storage diseases with their causal agents. This book provides a key knowledge on identification of all post-harvest diseases in a single volume with holistic approach on symptoms, epidemiological factors for the cause of the disease and management strategies which is supposed to be the first of its kind. This book comprehensively cover the post-harvest diseases of fruits and vegetables. It covers various titles like post-harvest diseases of fruits, vegetables, crops under protected cultivation and commodities under storage including cold storage methods.

I hope this book will serve as an instant guide to the Plant Pathology students, agricultural graduates and scientists and also the field level officers involved in both research and extension. I congratulate the author and contributors and I wish that the readers will reap the benefits by reading this book.

(N. KUMAR)

Date: 20.01.2021
Place: Coimbatore

• Tel : Off.: +91 422 2431788 *Res. : +91 422 2430887 *Fax: +91 422 2431672 *Email: vctnau@tnau.ac.in

Preface

The book entitled *"Postharvest Diseases of Perishables and its Management"* is a collection of chapters on postharvest diseases of fruits and vegetables prepared by the consortium of authors.

This book aims to provide comprehensive information to clearly understand the various kinds of symptoms produced by various postharvest pathogens causing postharvest diseases in fruits and vegetables and their management techniques.

It is hoped that simply designed this book definitely stimulate the teachers, researchers, microbiologist (Food microbiologist), Food technologists, Environmental scientists, Commercial production centers and Plant Quarantine and Certification agencies to focus the research on this aspects.

K. Prabakar

Contents

POSTHARVEST DISEASES OF VEGETABLES

Contents

POSTHARVEST DISEASES OF VEGETABLES

Chapter 1

Introduction

What is Postharvest Disease?

The diseases which develop on harvested parts of plants like seeds, fruits and also on vegetables are the postharvested diseases. The postharvest diseases that cause spoilage of both durable and perishable commodities are widespread. Greater losses occur in developing countries due to non-availability of proper storage and transportation facilities and improper handling methods, resulting in greater levels of injuries or wounds during harvesting and transit.

The harvested product might have been infected by pathogens prior to harvest under field conditions or they may get infected during transit and storage. The plant parts may get infected in the field, but expression of symptoms may take place later, at any stage before final consumption (latent infections).

The plant parts may get infected by micro-organisms and cause rotting or decaying – partially or totally. The quantity of plant products becomes reduced due to above infection.

The fleshy fruits, vegetables *etc.*, like tomato, banana, citrus, strawberries, rhizome of ginger, bulb of onion, tuber of potato *etc.*, may get damaged. This results in reduction of quantity or quality or both of the affected parts or products as a whole.

The amount of damage depends mainly on the pathogens involved, on the condition of the products and the condition of storage.

The pathogens involved are mainly fungi like *Pythium, Phytophthora, Rhizopus, Aspergillus etc.*, and some of the bacteria like *Pseudomonas, Erwinia etc.*

Classification of Postharvest Diseases

1. Field pathogens and
2. Storage pathogens

Field Pathogens

The field pathogens are those, which cause infection during development of plants or their products before harvest.

Storage Pathogens

The pathogens which cause infection during storage are storage pathogens.

Symptoms from infection caused by the "field pathogens" may be very inconspicuous to be noted at the time of harvest. In fleshy and or juicy fruits and vegetables, infection by field pathogen continues to develop even after harvest.

The fruits and vegetables may become infected during storage by the same field pathogens or by the other pathogens. In seeds and grains, the disease caused by field pathogens ceases to develop further soon after harvest. But, they may be infected further by the other pathogens during storage.

Importance of Postharvest Diseases

The postharvest diseases that cause spoilage of both durable and perishable commodities are widespread. Greater losses occur in developing countries due to non-availability of proper storage and transportation facilities and improper handling methods, resulting in greater levels of injuries or wounds during harvesting and transit. Durable commodities are generally stored in a dry state with moisture level below 12 per cent, whereas perishable products have higher moisture levels (about 50 per cent or more) at the time of storage. It is estimated that, in tropics, about 25 per cent of all perishable food crops harvested are lost between harvest and consumption. Losses in durable commodities, such as cereals, oilseeds, and pulses may be about 10 per cent on a worldwide basis. The losses may be both quantitative and qualitative. Under conditions favoring pathogens, loss caused by postharvest diseases may be greater than the economic gains achieved by improvements in primary production. Studies on postharvest diseases are primarily directed at preventing economic loss from spoilage of harvested commodities during transit and storage, and at eliminating the adverse effects of mycotoxins produced by fungal pathogens contaminating both durables and perishables. The mycotoxins are known to be carcinogenic, causing several serious ailments in humans and animals. There is an imperative need to gather information on the microbial pathogens involved in various postharvest diseases, conditions favoring disease development, and methods of developing effective systems of disease management.

Types of Postharvest Diseases

- Diseases of dry, bulk materials, such as seeds and grains, and
- Diseases of fleshy storage organs, such as vegetables (tubers, rhizomes, bulbs *etc.*) and fruits.

I. Diseases of Stored Seeds and Food Grains

- Field fungi, like *Alternaria, Fusarium, Cladosporium, Verticillium, Helminthosporium, Colletotrichum etc.*, attack seeds and grains on growing crops, but are unable to grow in storage due to low relative humidity *i.e.*, below 90 per cent.
- During storage or transit the seeds and grains are damaged by the different species of *Aspergillus* and *Penicilium,* which can grow well at a relative humidity ranges from 70-90 per cent. The commonly available *Aspergillus* species are *A. repens, A. ruber, A. flavus, A. candidus, etc.*
- *Aspergillus* and a number of other storage fungi invade the embryo of the seeds and grains and they discolour the embryo or seeds as a whole, thereby the germination percentage reduces markedly.
- In some cases, spoilage of stored grains and seeds results in drastic increase of temperature up to 70°C or more, which encourage the growth of different thermophilic and thermotolerant fungi such as *Aspergillus fumigatus, Absidia* spp., *Mucor pusillus, etc.*
- In addition to storage fungi, other microorganisms may grow in/or on seeds and accelerate the deterioration process. During breeding period of insects, the moisture content and temperature of seeds increase, thereby rapid growth of the pathogen takes place producing enormous amount of spores.
- During storage, the fungi produce mycotoxins that cause great damage to both domestic animals and human beings. The important fungi in this respect are *Aspergillus* and *Penicillium,* which produce aflatoxin and other toxins.

Most Important Toxins are

1. *Yellow Rice Toxins*

- Produced in grains of rice, barley *etc.*, by species of *Penicillium.*

2. *Tremorgenic Toxins*

- Produced on prepared food during storage in refrigeration or in other places and also on food produced from infected grains and/or seeds.

3. *Penicillic Acid*

- It is a carcinogenic substance produced by the different species of both *Aspergillus* and *Penicillium* in molded cereal grains.

II. Diseases of Vegetables and Fruits

- ✰ Different members of Ascomycotina cause the major postharvest diseases of fruits and vegetables.
- ✰ These are *Alternaria, Botrytis, Fusarium, Penicillium, Sclerotinia etc.*

1. *Alternaria*

- ✰ Different species of *Alternaria* cause rot of many fresh fruits and vegetables, *e.g.*, black rot of orange, tuber rot of potato, rot of sweet potato, purple blotch of onion, *Alternaria* rot of onion, *Alternaria* rot of cabbage, *etc.*

2. *Botrytis*

- ✰ It causes "gray mold rots" of fruits like pear, apple, citrus *etc.*, and vegetables like onion, tomato *etc.* Every year it causes great economic loss.

3. *Fusarium*

- ✰ It causes different diseases, commonly called "pink or yellow molds". Different species of *Fusarium* cause damage to tubers, bulbs, storage roots *etc.* and frequently on cucurbits, tomato *etc.* It also causes brown rot of fruits like lemon, orange *etc.*

4. *Penicillium*

- ✰ Species of *Penicillium* are commonly called "blue or green molds", these cause rots of different fruits like onion, sweet potato *etc.*
- ✰ They also cause spots on different fruits. Under storage, the spotted fruits bear tufts of spores.
- ✰ Though most of the *Penicillium* species prefer relatively high temperature for their growth in storage, they still remain active near freezing temperature - at a slow rate.
- ✰ A few species produce ethylene which increases respiration of fruits, thereby it reduces the storage life of the fruits.
- ✰ It also produces patulin - a mycotoxin - which directly contaminates the sauces and fruit juices prepared from infected partly rotten fruits.

5. *Sclerotinia*

- ✰ It infects different fruits and vegetables. Most common diseases are cottony rot of lemon, watery soft rot of bean pods, cucurbits *etc.*
- ✰ Storage diseases like bacterial soft rot of vegetables such as onion, carrot, potato *etc.* are mainly carried out by different species of *Erwinia,* such as *E. carotovora, E. chrysanthemi etc.*

Losses Due to Postharvest Diseases

Losses due to postharvest disease may occur at any time during postharvest handling, from harvest to consumption. When estimating postharvest disease losses, it is important to consider reductions in fruit quantity and quality, as some diseases may not render produce unsalable yet still reduce product value. For example, blemished fruit may not be sold as fresh vegetable but may still be suitable for processing, in which case, it brings a lower price. It is also important to take into account costs such as harvesting, packaging and transport when determining the value of produce lost as a result of postharvest wastage. Aside from direct economic considerations, diseased produce poses a potential health risk. A number of fungal genera such as *Penicillium*, *Alternaria* and *Fusarium* are known to produce mycotoxins under certain conditions. Generally speaking, the greatest risk of mycotoxin contamination occurs when diseased produce is used in the production of processed food or animal feed. In most cases, fresh produce which is obviously diseased would not be consumed. Losses due to postharvest disease are affected by a great number of factors including:

- ✰ Commodity type.
- ✰ Cultivar susceptibility to postharvest disease.
- ✰ The postharvest environment (temperature, relative humidity, atmosphere composition, *etc.*)
- ✰ Produce maturity and ripeness stage.
- ✰ Treatments used for disease control.
- ✰ Produce handling methods.
- ✰ Postharvest hygiene.

Virtually all postharvest diseases of vegetables are caused by fungi and bacteria. In some root crops and brassicas, viral infections present before harvest can sometimes develop more rapidly after harvest. In general, however, viruses are not an important cause of postharvest disease. Postharvest diseases are often classified according to how infection is initiated. The so-called 'quiescent' or 'latent' infections are those where the pathogen initiates infection of the host at some point in time (usually before harvest), but then enters a period of inactivity or dormancy until the physiological status of the host tissue changes in such a way that infection can proceed. The dramatic physiological changes which occur during fruit ripening are often the trigger for reactivation of quiescent infections. Examples of postharvest diseases arising from quiescent infections include anthracnose of various tropical vegetables caused by *Colletotrichum* spp. and gray mould of cucurbits caused by *Botrytis cinerea*. The other major group of postharvest diseases is those which arise from infections initiated during and after harvest. Often these infections occur through surface wounds created by mechanical or insect injury. Wounds need not be large for infection to take place and in many cases may be microscopic in size. Common postharvest diseases resulting from wound infections include blue and green mould (caused by *Penicillium* spp.) and transit rot (caused by *Rhizopus stolonifer*). Bacteria such as *Erwinia carotovora* (soft rot) are also common wound invaders. Many pathogens, such as the banana crown rot fungi, also gain entry through the injury created by severing the crop from the plant.

A wide range of microorganisms attack vegetables during postharvest period. Among different pathogenic microorganisms, fungi are the major organisms deteriorating fruits after harvest. Bacterial pathogens are not common in fruits due to their acidic nature. Postharvest diseases of some important crops are discussed below.

Postharvest Diseases of Fruits

Chapter 2

Postharvest Diseases of Mango and their Management

K. Prabakar

Department of Plant Pathology, Centre for Plant Protection Studies, Tamil Nadu Agricultural University, Coimbatore – 3, Tamil Nadu

Mango (*Mangifera indica*) belongs to a family Anacardiaceae and it is recognised as King of fruits. It is one of the delicious fruit having minerals, vitamins and nutritious qualities. It is usually grown in tropical and subtropical regions of the world including subcontinent. It is an important fruit hence plays an important role in the economy via export. However, mango is susceptible to a number of diseases at all stages of its development from the seedling to the fruits. The mango tree and fruit is the usually host of different pathogens specifically fungi which produces postharvest fruit rot worldwide. Although mango is affected by large number of diseases but some diseases are of great economic importance and are responsible for high loss in the mango production in our country.

Anthracnose

Pathogen: *Colletotrichum gloeosporioides*

Symptoms

- Infection occurs leaves, stems, young flowers and fruit
- Sunken black spots appear on the surface of the fruit during ripening.
- The anthracnose symptoms in the fruit results from latent infection which is early stage of the infection in immature fruit due to antifungal

compounds present in the peel. During repining stage, the quiescent pathogen is causing anthracnose symptoms as shown in the figure.

Sunken Black Spots

***Courtesy*: TNAU Portal**

Management

Pre-harvest Control

- ✰ Spray with Mancozeb (800 g/kg at 2 g/L) weekly during flowering and then monthly until harvest.
- ✰ Stop spraying 14 days before harvest.
- ✰ During dry weather, flower sprays may be reduced to fortnightly intervals.
- ✰ Should rain occur during flowering, apply with Mancozeb1 g product/L.
- ✰ Copper oxy-chloride sprays (4 g/L) used for bacterial black spot control also control anthracnose, however Copper oxychloride should not be used during flowering.
- ✰ Where bacterial black spot is serious, Copper oxychloride can be substituted for Mancozeb sprays after flowering.
- ✰ Spraying of *Bacillus pumilis* strain 2808 @1.38 per cent to arrest pathogen during latent infection

Postharvest Control

- ✰ Dip fruits within 24 hours of harvest by totally submerging them for 5 minutes in hot water (52°C). Lowering the temperature of the dip below 52°C will reduce the effectiveness of the treatment.
- ✰ With an approximate ratio of 3 litres of dip to 1 kg of fruit, no appreciable temperature drop occurs when fruit is added to the tank. This dip also partially controls stem end rot.

Stem End Rot

Pathogen: *Lasiodiplodia theobromae*

Symptoms

- In fruits, the pericarp darkens near the base of the pedicel.
- The affected area enlarges to form a circular, black patch which under humid atmosphere extends rapidly and turns the whole fruit completely black within two or three days.
- The pulp becomes brown and softer.

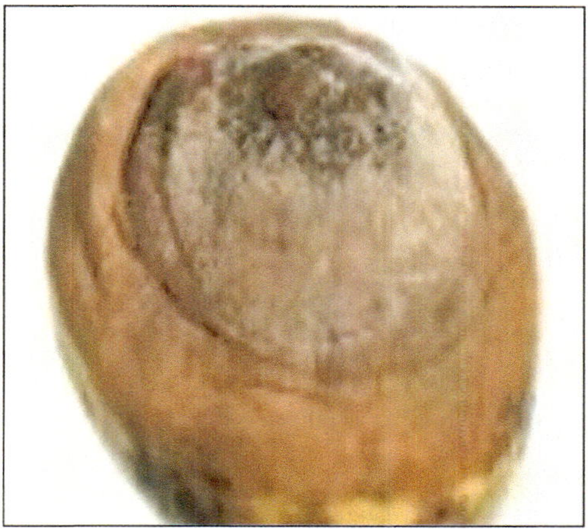

***Courtesy*: TNAU Portal**

Management

- Avoid harvesting immature fruit.
- Postharvest treatment with hot water and Carbendazim is partially effective against stem end rot.
- For control of stem end rot during controlled atmosphere storage, a dual treatment of hot Carbendazim followed by Prochloraz is necessary.
- Fruit from orchards with a history of stem end rot losses should be rejected for long term storage.
- The severity of stem end rot can be assessed as follows.
- Harvest 100 mature fruit at random from throughout the orchard.
- Leave them untreated and store at 25°C until they fully ripe.
- Ideally, less than one-tenth and certainly not more than one-third of the fruit should develop symptoms of stem end rot by the time they are fully ripe.

Black Mould Rot

Pathogen: *Aspergillus niger*

Symptoms

- Infection occurs during and after harvest, when fungus penetrates through wounds or the cut ends.
- The infection usually starts at injured portion of the fruits or stem end and develops into grayish to pale brown spots mainly from stem end portion, coalescing into dark brown to black lesions and later covered with a sooty mass of brown/black spores.
- The spots are soft and sunken, spreading rapidly.
- In advance stages copious growth of fungal colonies develops.

***Courtesy*: W.M. Haggag**

Management

- Careful handling of fruits at all stages to avoid mechanical injury and sap bum damage to the fruits.
- Preharvest sprays and postharvest dip of fruit in hot water containing Carbendazim (0.05 per cent) for 5m at 52± 1°C to prevent fungus already seated in the injuries.
- Storing fruits at 12°C prevents the rotting.

Botryosphaeria Rot

Pathogen: *Botryosphaeria ribis*

Symptoms

- The first sign is a slight softening of the tissue around the spot.
- The affected skin turns brown and the flesh beneath may darken. Infection may also occur via injuries or any part of the skin.
- Lesions are dark, sunken and oval to round.
- Elliptical with a shallow decay of the flesh beneath.
- Minute black bodies are formed in the skin and white to gray mouldy growth is developed.
- Various symptoms *viz.,* purplish brown spots and stem end rot have been noticed on mango fruits with a decay of flesh beneath.
- Because of the variability in symptoms produced by fungus, identification necessitates culturing of causal fungus and examining its spores microscopically.

Management

- Strict orchard hygiene and good tree vigour should be maintained.
- Pre-harvest spraying of fungicide *viz.*, carbendazim (0 1 per cent) and dipping of fruits in hot water (52±1°C for 30m) after harvest.
- Harvesting of fruits with stalk is recommended to avoid infection.

Brown Rot

Pathogen: *Lasiodiplodia theobromae*

Symptoms

- The disease mostly appears from the stalk or at anywhere in wounded fruit.

***Courtesy*: W.M. Haggag**

- It is characterized by dark brown to black necrotic patches not clearly defined, margin ultimately encircling the entire fruit.
- Development of abundant small black pycnidia is observed on decayed fruits.
- The infection is noticed on ripened fruit but absent on green ones.
- A heavily infected fruit turn dark, internal pulp disintegrates, becomes brittle and light in weight.

Management

- Careful handling of fruit to avoid bruising/wounding.
- Pre-harvest sprays of Carbendazim or Thiophanate methyl (0.1 per cent) at 15 days intervals.
- Hot water dips of harvested fruits.

Stem End Soft Rot

Pathogen: *Ceratocystis paradoxa*

Symptoms

- The fungus has been observed in Indian mangoes.
- It induces a soft rot and causes extensive postharvest losses during handling of fruits.

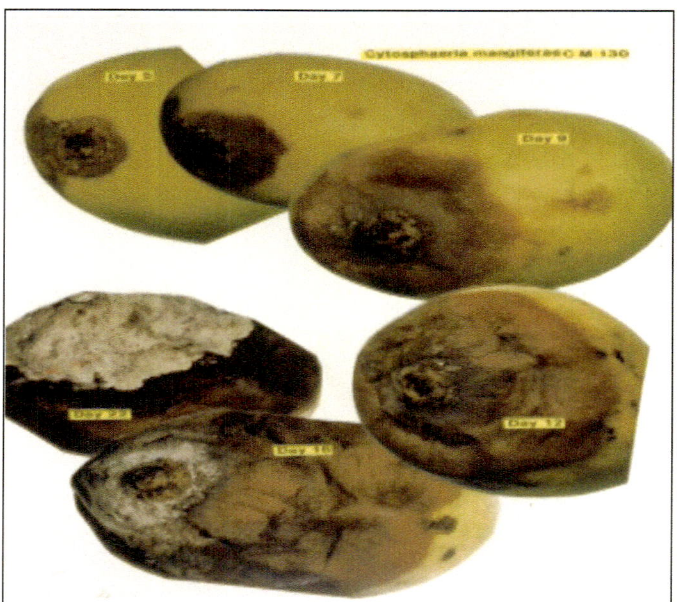

***Courtesy*: W.M. Haggag**

- Infection occurs only in wounded fruits around the stem end as pale, soft and watery, irregular lesion, but later darkens with the formation of spores and subsequent black spore bearing bodies are seen on the lesions.
- Lesions gradually spread covering the entire fruit.
- Rotting is accompanied by a rancid odour.

Management

- Hygienic practices are beneficial in reducing losses.
- Affected fruits should be promptly destroyed.
- Avoid picking and packing injuries, sun burn or wet fruits while handling.
- Fruits should be treated with metalaxil @0.7g/lit to prevent postharvest infection.

Charcoal Rot

Pathogen: *Macrophomina phaseolina*

Symptoms

- It is characterized by black irregular necrotic leathery patches usually originating from distal end of the fruits.
- The fungus usually infects through the cut ends.
- The tissue starts rotting from below and turns brown with numerous black irregular to oval pycnidia and eventually the entire fruit darkens.
- The internal pulp gradually turns dark brown and ultimately rots emitting foul odour under humid conditions.

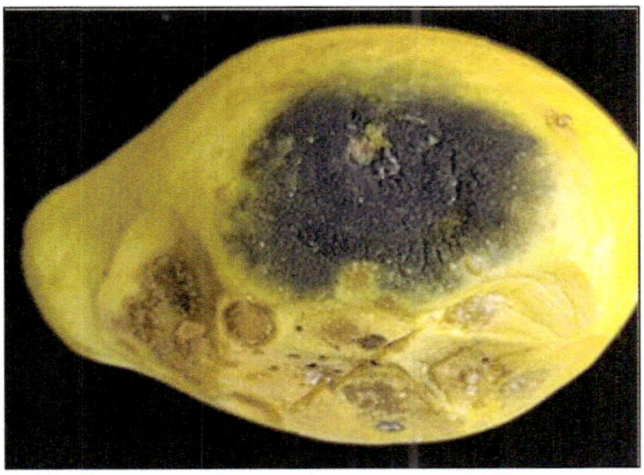

***Courtesy*: S. Dhawan**

Management

- Field hygiene is very important, and the removal of diseased fruits helps to limit the disease spread before sclerotia can proliferate and fall on to the soil.

- ☆ Careful handling of harvested fruits to avoid soil contamination and bruising.
- ☆ Fruits, which may have field infection, can be treated by means of a postharvest fungicide dip.

Alternaria Rot

Pathogen: *Alternaria tenussima* and *Alternaria alternata*

Symptoms

- ☆ On fruits, initially the spots are water soaked and later become black, circular and sub-cuticular.
- ☆ It gradually enlarges and become irregular.
- ☆ At first, spots are concentrated around the extreme end of the fruit where high numbers of lenticels are present.
- ☆ Initially the decay is firm and does not penetrate into the pulp.
- ☆ Later, the disease makes headway into the flesh.
- ☆ The centre of the spots is slightly sunken.
- ☆ Under humid conditions dark brown spores are seen over the lesions.
- ☆ In advance stages, the whole fruit becomes rotten.
- ☆ When the skin of diseased fruit is removed, reddish patches can be seen on the flesh below the lesions.

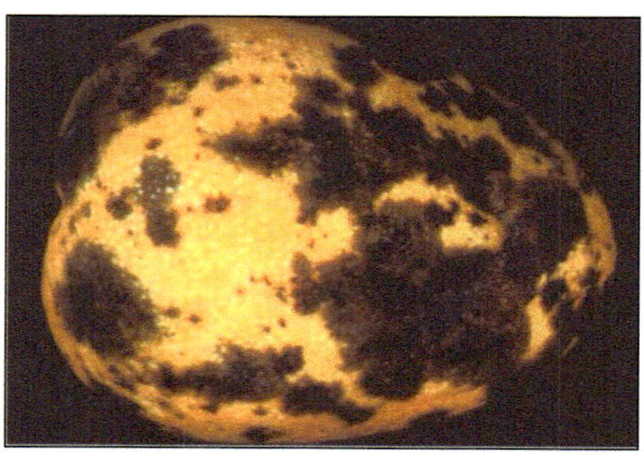

Courtesy: **Tom A. Zitter**

Management

- ☆ Preharvest treatments reduce the latent infection, whereas postharvest treatments inhibit the development of latent infection.

- Sprays with protectant fungicide Mancozeb (0.2 per cent) at 15 days interval until fruit set.
- Postharvest dip with Thiophanate methyl (0.1 per cent) in hot water is recommended.
- Fruits should be stored at lower temperature (10-12°C) and marketed promptly.

Phoma Rot

Pathogen: *Phoma mangiferae*

Symptoms

- The disease is noticed in Dashehari and Banganpalli varieties.
- Black to brown spots are present on the entire fruit surface.
- Outer skin is hard but the internal portion rot quickly.
- In advance stage, dark pycnidia appear over such rotten tissues.

***Courtesy*: Kelly Ivors**

Management

- Careful handling and storage of fruits at lower temperature should be done.
- Foliage, fallen from the tree as primary source of infection and harbouring the pathogen, should be burnt.

Rhizopus Rot

Pathogen: *Rhizopus arrhizus*

Symptoms

- Generally the fungus enters the fruit through injuries and becomes more destructive at higher temperature.

- The disease is characterized by large water soaked areas developing on any portion of the fruit with profuse growth of the fungus with black sporangial bodies.
- Skin is easily displaced from infected flesh.
- This type of soft decay is very common, emitting foul smell.
- There is profuse development of coarse white mould strands giving rise to globular white spore heads (sporangia), which later turn black and easily visible to naked eyes.
- They arise from dark branched, root-like structure (rhizoids), an exception is a form of *R. oryzae* known as *R. arrhizus*, which does not possess rhizoids.

***Courtesy*: R.C. Ploetz**

Management

- Strict hygiene in orchard is of utmost importance.
- Pre-harvest sprays help to reduce postharvest rotting.
- Fallen and rejected fruits should be removed and destroyed.
- Care should be taken during harvesting and handling of the fruits.
- Postharvest dip in Thiophanate methyl (0.1 per cent) and heat treatment (52±1°C for 30m) is essential to deactivate latent infection.
- Do not hold the fruit at high temperature.
- Fruits should be pre-cooled after treatment and stored at lower temperature (10-12°C) after fungicidal treatment.

Mucor Rot or Leak

Pathogen: *Mucor subtilissimus*

Symptoms

- It is essentially a wound parasite.
- Strands of white mould develop on the surface of infected fruits.
- Tile fungus may cause a lateral rot or more; usually a stem end rot or at any place on the injured fruit surface.
- Water soaked lesions become covered with yellowish mould bearing black spore-heads (sporangia).
- The decay is accompanied by yeasty odour.
- Infection is very high when skin is punctured during egg laying of fruit flies contaminated with spores.
- After few days when the temperature rises to more than 20°C the entire fruit is decayed accompanied by unpleasant smelling.
- Fungus is capable of breaking down the fruit tissue so that juice leaks out, hence the common name for the disease.

***Courtesy*: J.A.Von Arx**

Management

- Fallen fruits should be promptly collected and destroyed.
- Harvesting of fruits may be avoided in wet weather.
- Avoid skin injuries/bruises.
- Boxes must be cleaned to remove soil.
- Chlorination is beneficial to control the disease.
- Dipping of fruits in fungicide *viz.*, Carbendazim (0.1 per cent) solution with sticker/spreader.

Gray Mould

Pathogen: *Botrytis cinerea*

Symptoms

- A light brown firm decay develops from the stem end or from wounds affecting a large portion of the fruit.
- Gray or brown fungal growth containing mass of spores covers the diseased surface.
- The pathogen spreads readily by contact with adjacent fruit, giving rise to larger nests of diseased fruit in packed containers.

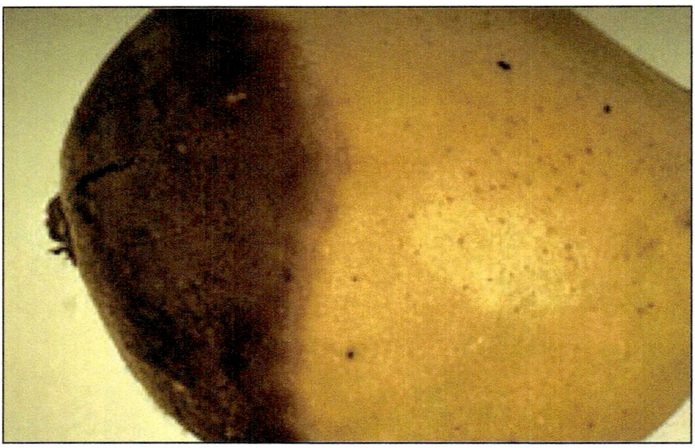

Courtesy: **M.W. Alam**

Management

- Dip fruit within 24 hours of harvest for five minutes in hot water (52°C) having 100 g Benomyl per 1 lit of water.
- Since soil is a major reservoir of the fungus, fruit lying on the surface should not be included in the harvest.
- Injuries to fruit during harvesting and handling should be minimized.

Blue Mould

Pathogen: *Penicillum expansum*

Symptoms

- A soft brown decay of the skin and flesh develops at the stem end or at wound and may spread on the entire fruit.

- White powdery spore masses, which become blue when mature, may develop on the stem end, at wound site or in internal cavities at the stem end.

***Courtesy*: Tom A. Zitter**

Management

- Application of Benomyl (0.1 per cent) or Prochlorez (0.1 per cent) provide control to some extent.

Chapter 3

Postharvest Diseases of Banana and their Management

S. Parthasarathy and K. Prabakar

Department of Plant Pathology, Centre for Plant Protection Studies, Tamil Nadu Agricultural University, Coimbatore – 3, Tamil Nadu

India is progressing towards self-sufficient in fruit and vegetables. The industry is intensive, typically seasonal in operation and dominated by small-scale farms. Most of our production is for domestic markets, but the horticulture industries also contribute significantly to exports. Fruit are among the most important and interesting food crops that are produced in the tropical regions of the world. Pests and diseases are major constraints in horticultural production system. The conventional pest and disease control is mainly based on application of fungicides and resistance breeding. The resistance breeding mainly depends on the availability of resistance genes, which often show limited durability. On the other hand, indiscriminate use of pesticides for the control of crop pests and diseases has led to many problems in soil, water, air, food and fodder. Thus, it has become mandatory to develop alternate technologies that reduce the use of pesticides in pest and disease management. Effective, integrated disease management and prevention strategies have been essential for minimising losses and facilitating market access for nation development as a major producer and exporter of high-quality fruit and vegetables. Among fruit species grown in India, Banana is major fruit commodities with potential for local consumption, export and import substitution. Since, increase in area of cultivation is impossible, the alternative approach is to increase the productivity, but the main hurdle in increasing the productivity is the threat posed by pests and diseases.

Banana and plantain is a dessert fruit or a starch staple to a large population of the world. The fruit is rich in vitamins A, B, C, minerals and energy. Several fungi are reported to cause postharvest losses in banana. Reduction in total sugars, total soluble solids and vitamin C in fruits infected with *Colletotrichum gloeosporoides*. Under pathogenesis, changes in protein and phenol contents in banana-fruits were also recorded. Protein concentration exhibited gradual fall under pathogenesis by prominent fungi, *Botryodiplodia theobromae, Diplodia natalensis, Fusarium oxysporum, Helminthosporium speciferum, Curvularia lunata, Aspergillus flavus* and *Trichothecium roseum*. The maximum reduction was observed due to *B. theobromae*. Fungal association and changes in carbohydrate and protein contents of fruits have been well understood. Fungal population associated with fruits increases as ripening proceeds. Carbohydrates and proteins increase at the initial stages of ripening but decline after 12 days. Calcium carbide application hastened ripening process in cv. Nendran as carbohydrate and protein contents decreased by eighth or nineth day after treatment. Ripening is correlated with fungal association and reduction in shelf-life of fruits. Gradual decline in total phenols, orthodihydric phenol and anthocyanin contents of banana-peels also reduce during rotting, and thus pre-dispose fruits to fungal infection.Here, the postharvest diseases of national economic importance are discussed with their latest management methods.

Anthracnose

Pathogen: *Colletotrichum musae*

Symptoms

- There are two types of symptom, resulting from different modes of infection.
- Young latent infections appear as small black-and-circular specks on the flowers, fruit-peels and distal-ends of banana-hands.
- Typical anthracnose lesions on green fruit are generally dark brown to black with a pale margin, lenticular in shape, slightly sunken and with dimensions of several cm.
- On ripened fruits the typical symptoms are numerous small dark circular spots which enlarge, coalesce and become sunken.
- The lenticular lesions may also be present in ripening fruit.
- On both types of lesion salmon-pink spore masses are eventually produced.
- Severely infected fruits become dark due to blemishes.
- Infected fruits prematurely ripen and turn black and rotten.

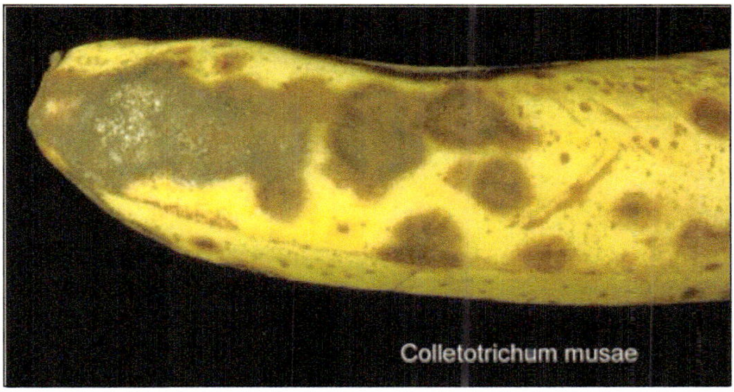

Courtesy: **Don Edwards**

Management

- Protection from latent infection can be achieved through fungicidal spraying.
- Spraying bunches once a month with copper fungicides until fruit is ready for cutting has been found effective.
- Dithiocarbamates are also found useful in controlling diseases.
- Extracts of *Solanum torvum* was reported to be more effective in controlling the banana anthracnose disease caused by *Colletotrichum musae*.
- The effect of systemic fungicide Benomyl in controlling established latent infection by *C. musae.*
- Cutting off hands from bunches and packing them in cardboard boxes have minimized incidence of anthracnose in bananas to a great extent.

Crown-Rot

Pathogen

Botryodiplodia theobromae Pat, *Cephalosporium* sp., *Ceratocystis paradoxa* (Dade) Moreau., *Colletotrichum musae* (Burk and Curt.) Arx., *Fusarium pallidoroseum* (Cooke) Sacc., *Acremonium* spp. and *Verticillium thebromae* (Turc.) Mason and Huges., *Musicillium theobromae, Lasiodoplodia theobromae, Nigrospora sphaerica, Penicillium* spp. and *Aspergillus* spp., as well as many *Fusarium* species including *F. semitectum, F. verticillioides, F. sporotrichoides, F. oxysporum* and *F. solani.*

Symptoms

Lasiodiplodia theobromae

- It is a wound parasite, it readily attacks cushions after they have been dehanded.

- Infection extends from cushion region to finger-stalks, and from there to fingers, resulting in rotting and dropping of individual fingers.
- Along with this, a brownish black discolouration is also noticed on the surface of the fruit, and the fruit becomes soft and wrinkled.
- Light-gray cottony mycelium covers the whole fruit.
- Later peel gets shriveled and encrusted with pycnidial mass.
- Maximum damage to fruits occurs at 23-30°C and minimum at 7°C.

Ceratocystis paradoxa

- Infection occurs when main stalk is cut or bananas are dehanded.
- Infected main stalks decay rapidly; tissues become blackened and soft and give a characteristic sweetish odour.
- Under humid conditions, fungus forms a cobweb-like mass of dark-coloured mycelia that invade cushion, finger-stalks and fingers.
- Infected fruits show uneven black discolouration of peel, and pulp is reduced to soft-and-wet consistency of dark-brown colour.

Colletotrichum musae

- Infection occurs through small peel injuries on immature banana-fruits and results in the development of lesions; as detailed under anthracnose.
- In the ripening room, *Colletotrichum* infection also produces tip-rot when fruits are turning yellow.

***Courtesy*: S. Nelson**

Management

- In areas where fruit diseases cause serious losses, direct spraying of individual bunches (after 10-20 days of emergence) with fungicides is required.

- Three to four sprayings with any of the following chemicals can be given: Zineb (0.2 per cent) or Maneb (0.2 per cent).
- Systemic fungicides also offer protection. Bavistin (500 ppm), Benomyl (500 ppm) and Thiabendazole (200-400 ppm).
- Storage at 7°C increases shelf-life of bananas up to 10 days without much changes in fruit quality.
- Yeast *Candida oleophila* suspension applied @ 10^7 cfu/ml with 2 per cent $CaCl_2$, followed by storage under modified atmospheric packaging (MAP) in the non-perforated polyethylene bags reduced crown-rot significantly as compared to untreated control.

Speckle/Black-Spot/Black-Tip

Pathogen: *Deightoniella torulosa*

Symptoms

- The fungus causes symptoms on banana leaves and fruits.
- The disease is characterized by black discolouration just below the flower that progresses along the fruit.
- The diseased area is bordered by an irregular narrow gray or pale yellow margin.
- One side of the fruit is affected more than the other.
- Larger spots may measure up to 4 mm in diameter with a sunken centre, which may crack as fruit matures.

***Courtesy*: W.S. Nishijima**

Management

- Black tip can be controlled by improving hygiene and drainage in the plantation.

Pitting Disease/Johnston-Spot

Pathogen: *Magnaporthe grisea, Pyricularia grisea*

Symptoms

- They appear as small reddish spots on the green-fruit peel, which is approaching maturity.
- Later, these spots develop into typical small, shallow black pits, which may affect the bunch appearance.
- Sometimes no apparent symptoms are noticed on the harvested bunch, but pits may develop during transport and storage.
- In the ripening room, fungal mycelia develop on the surface of these spots, and as the ripening proceeds, original pits get overgrown by secondary saprophytes.

Courtesy: **S. Nelson**

Management

- Preharvest spraying with any of Dithio-carbamates at 0.2 per cent concentration when bunch begins to develop gives protection.

Cigar-End Rot/Banana Fruit/Rot

Pathogen: *Verticillium theobromae*

Symptoms

- One or all fingers may be affected by a tip-end rot which starts with localised darkening and wrinkling of the skin.

- These appear from the tip of the immature fruits at the pistillate-end and spread upwards.
- The peel becomes folded and shrunken as infection spreads slowly along the finger.
- Ashy conidia and conidiophores cover rotten portions, which become corrugated and impart a burnt ashy cigar appearance with a dark border.
- Decay may extend up to one-third of the fruit, and the fruit develops a typical dry-rot.
- In *Trachysphaera* rot the surface of the lesion becomes covered with white spores which later turn pink or brown; internally the pulp may undergo a wet rot or, in the absence of secondary organisms, may become dry and mummified.
- In *Verticillium* rot the tissue is characteristically dry and fibrous, and the spores are gray and powdery.
- In both diseases the symptoms bear a resemblance to the ashy end of a burnt cigar.

***Courtesy*: Don Edwards**

Management

- Spraying of bunches at the time of flowering (after all hands open completely) with Methyl- 2-benzimi-dazole carbamate (MBC) (0.1 per cent), Benomyl (500 ppm) and Mancozeb (0.2 per cent) is effective in reducing cigar- end rot.

Squirter Disease

Pathogen: *Nigrospora sphaerica*

Symptoms

- ✰ The fungus penetrates through broken stem-ends resulting from mechanical injury, and converts pulp into a dark watery fluid without any symptoms on the peel.

***Courtesy*: Don Edwards**

Management

- ✰ Dipping the hands in Thiabendazole solution of 0.01 per cent was promising.

Freckle/Black-Spot Disease

Pathogen: *Macrophoma musae* (Cooke) Berl. and Vogl. *Guignardia musae* Racib. (conidial state: *Phyllosticta musarum* (Cooke) van der Aa).

Symptoms

- ✰ The disease is characterized by spotting on the leaves and fruits.
- ✰ The fungus develops numerous minute, grayish brown or dark brown, raised, more or less rounded spots, 1 mm in size with black dots in the centre.
- ✰ Lesions are present in abundance on the upper surface of the fruits with a narrow water-soaked margin.
- ✰ Minute projecting pycnidia render roughness to fruits.
- ✰ In case of severe infection, affected fruits appear blackish.

Management

- ✰ Control measures are not necessary as the symptoms are superficial; on the peel only.

***Courtesy*: Don Edwards**

Fusarium Tip-Rot

Pathogen: *F. moniliforme*

Symptoms

- The disease is characterized by tip rotting of the young immature fruits.
- The infection originates from the base of the decaying perianth and causes blackening and wrinkling of peel, and premature yellowing of fruits.
- In advanced stage, brown-peel gets covered with rosy powdery fungus growth and inner pulp becomes soft and rotten.
- Symptoms due to *F. moniliforme* are evident on cutting fruits longitudinally.
- Deep-brown discolouration extending from tip to bottom is noticed; tissues become soft and pulpy.

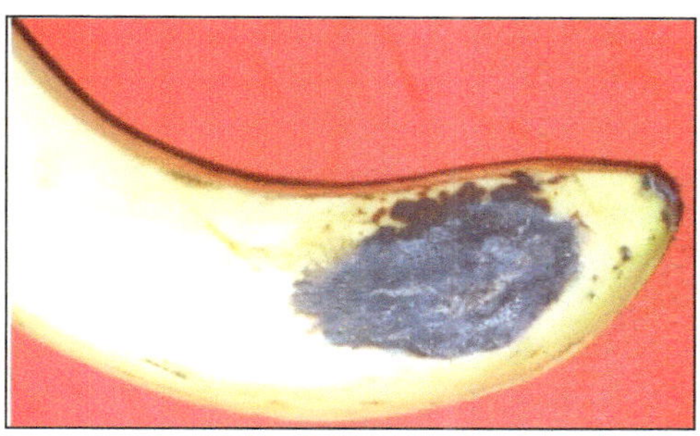

***Courtesy*: Don Edwards**

Management

- Careful handling is essential to avoid wounds and bruises during transport and storage.

- Storage of fruits at or below 10°C and relative humidity at 25 per cent is ideal for preventing storage-rots.
- Aureofungin has been reported to be effective.

Rhizopus Rot

Pathogen: *Rhizopus stolonifer* (Ehrenb.) Vuill., *R. oryzae* Went and Prinsen Geerligs., *R. arrhizus* Fischer

Symptoms

- They appear as brown irregular lesions, which gradually increase in size, and within a short span, entire fruit surface gets covered with fungal growth.
- These results in softening of fruits, and decaying fruits emit a fermented odour.

Management

- Collect bunches during the dry period from January to april (Kerala and Tamil Nadu conditions) could be stored for longer period without fear of microbial spoilage as compared to bunches collected during other periods of the year.
- Storage needs to be carefully done, and rapid cooling is required.

Green Mould-Rot

Pathogen: *Aspergillus flavus* Link, ex Fries

Symptoms

- This rot develops as water-soaked lesions on fruits, which later become soft and juicy.
- In advanced stages, fruit-surface gets covered with olive-green dot-like mycelia with abundant sporulation.

***Courtesy*: Don Edwards**

Management

- Avoid injuries and bruises on fruits during handling and storage.
- Fruits decay can be controlled by dipping them in any of the following fungicides/chemicals, Thiophenate methyl, Benlate, Thiabendazole, Carbendazim, Propionic acid, sodium metabisulphite (at 2,000 ppm).
- The shelf-life could also be extended by 8 days.

Black Mould-Rot

Pathogen: *Aspergillus niger*

Symptoms

- The disease manifests itself as water-soaked circular lesions around the injured tissues.
- Rotting is gradual and whole surface gets covered with black fungal growth, and ultimately dry-rot symptoms appear.
- *Aspergillus oryzae* also caused similar symptoms.
- In advanced stages, affected fruits become black and shrivelled.

Management

- Bunches should be healthy while they are stored and ripened. Care should be taken as follows
- Harvesting of bunches sholud at the correct stage of maturity (over – ripening should be avoided)
- Rapid cooling is required.

Watery Soft-Rot/*Sclerotinia* Fruit rot

Pathogen: *Sclerotinia sclerotiorum*

Symptoms

- Infection starts from the perianth region and progresses along the fingers with a brownish water-soaked appearance, which later turns dark-brown.
- Internal rotting is also noticed.
- The white inner tissue becomes reddish-black, and finally gets broken down and hollowed out.
- Black seed-like bodies representing fungal sclerotia are produced in abundance.

Management

- Storage needs to be carefully done, and rapid cooling is required.
- Introduction of rapid transport to distant markets with adequate facilities for controlling temperature helps reducing incidence of many diseases.

Courtesy: **Don Edwards**

Botrytis End-Rot

Pathogen: *Botrytis cinerea*

Symptoms

- The fungus produces a wet-rot.
- Infection begins at the fruit-tip or perianth, and spreads causing dark-brown discolouration on the peel, followed by rotting of the pulp.
- The surface of the decayed fruit is covered by dense, brown mycelia, characterized by sticky, wet, dark-brown epicarp and a reddish-green endocarp due to mycelial growth and fruiting bodies.

Management

- In the collecting stations, holds and storage rooms, cleanliness is must to minimize contamination.
- Bunches sholud be healthy while they are stored and ripened.
- Care should be taken while handling the bunches.

Pink Mold-Rot

Pathogen: *Trichothecium roseum*

Symptoms

- It originates mostly from stigma-end, and appears as pale-brown water-soaked lesions that enlarge to form large patches with a pink moldy overgrowth.

Management

- Handle bunches with care to minimize bruising.

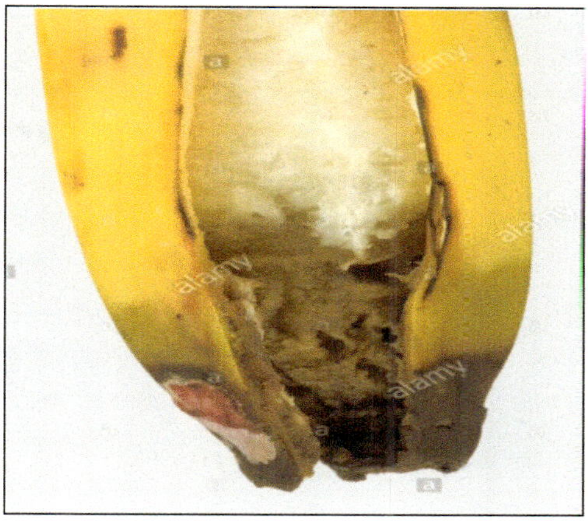

***Courtesy*: Don Edwards**

Helminthosporium Fruit rot

Pathogen: *Helminthosporium spiciferum*

Symptoms

- ✰ This disease caused considerable rotting to banana-fruits.
- ✰ In fruits, gradual loss of ascorbic acid (Vit. C) content was also noticed; due to production of ascorbic-acid degenerating enzymes, either by fungus or due to host-pathogen interaction.

Management

- ✰ Cleanliness is a must to minimize contamination.
- ✰ Storage needs to be carefully done.

Dothiorella-Rot

Pathogen: *Botryosphaeria ribis*

Symptoms

- ✰ Decay begins characteristically from the tip-end of the infected flower part and develops as in the finger-rot.
- ✰ Minute black bodies (pycnidia) are formed on the peel.
- ✰ A white-to-gray mould growth may also develop.

Management

- ✰ Harvesting of bunches sholud at the correct stage of maturity (over – ripening should be avoided)

Courtesy: **Don Edwards**

- Bunches should be healthy while they are stored and ripened.

Banana Black-Rot

Pathogen: *Thielaviopsis paradoxa* Van. Honn

Symptoms

- Fruits become soft and black.
- Under in-vitro conditions, intensity of the rotting was found maximum at 35°C and at 100 per cent relative humidity.

Courtesy: **Don Edwards**

Management

- Storage of banana fruits at 10°C and below 25 per cent RH is effective in reducing spoilage by *T. paradoxa* without affecting fruit quality.

Sooty Mould

Pathogen: *Capnodium* sp.

Symptoms

- The mould is associated with the honey-dew secreted by banana-aphid. Occasionally fungus *Cladosporium* sp. was also found associated with it.
- Small, dark-brown flecks appear on the crown, pedicels and fingers of the fruits. These flecks increase in size and merge into blackish brown blotches that may extend up to $1/3^{rd}$ of the fruit.

***Courtesy*: Don Edwards**

Management

- The disease can be controlled by dipping in bleaching solution (0.05-0.1 per cent).

Sigatoka

Pathogen: *Mycosphaerella fijiensis* Morelet

Symptoms

- Fruit symptoms include buff-coloured pulp, premature ripening, increased susceptibility to chilling injury (q.v.), and development of abnormal flavour and aroma.
- Produces lesions on the leaf margins known as 'tip-spotting'.
- *M. fijiensis* causes blackleaf streak, in which especially severe symptoms result from ascospore infection, while *M fijiensis* var. *difformis* causes black sigatoka, with similar symptoms.

Courtesy: **Don Edwards**

Management

- The disease can be controlled by dipping in bleaching solution (0.05-0.1 per cent).

Fruit Rot

Pathogen: *Penicillium citrinum*

Symptoms

- Diseased fruit's skin becomes soft and shows numerous green pinhead like structures.
- On the removal of peel, pulp appears soft, yellow and shrivelled with water-soaked appearance.
- Fungi population increases with ripening is significantly higher in fruits where ripening is hastened with calcium carbide.

Curvularia Fruit Rot

Pathogen: *Curvularia lunata*

Symptoms

- The symptoms appear as water-soaked sunken lesions with irregular margins.
- In the beginning infection is limited to peel.
- But in advanced cases, infection occurs in the pulp also.

Management

- Dipping fruits in propionic acid or salicylic acid or sodium metabisulphite (at 2,000 ppm) saved banana fruits from infection.

Fruit Rot

Pathogen: *Chaetosphaeropsis truncata*

Symptoms

- Infection occurs at the stylar-end of the fruits as brownish and water-soaked skin discolourations; spreading around the point of incidence with rotting spreading upwards and covering fruit within 8-10 days.
- Skin colour turns black.
- In severe cases, skin shrivels and numerous dark-coloured pycnidia of pathogen appear all-over the fruit surface.
- Pulp rotting is faster than that of skin.

Chapter 4

Postharvest Diseases of Grapes and their Management

V. Senthilvel, S. Vanitha and K. Prabakar

Department of Plant Pathology, Centre for Plant Protection Studies, Tamil Nadu Agricultural University, Coimbatore – 3, Tamil Nadu

Grape is one of the most diffuse fruit in the world both as fresh fruit (table grape) and processed in wine, grape juice, molasses, and raisins. The reason to have these different processed products depends on the extreme perishability of the fruit. As fresh fruit, grapes are very delicate and the loss at harvest and during the distribution is very high. The quality of grapes in market not only depends on various activities carried out in the vineyard, but the operations and handling during and after harvesting also play important role. As grape is non-climacteric fruit, should be harvested when they are fully ripen since neither the colour nor the taste improves after harvest. Harvesting period is determined by the variety, climatic conditions, TSS, acidity and sugar acid ratio, depending on whether the grape is for local or export market.

Anthracnose/Bird Eye Spot

Pathogen: *Elsinoe ampelina*

Symptoms

- ✰ On berries, small, reddish circular spots initially develop.
- ✰ The centers of the spots turn whitish gray and are surrounded by narrow reddish-brown to black margins.

- Acervuli (fungal fruiting structures) eventually develop in the lesions.
- A pinkish mass of fungal spores (conidia) exudes from these structures during prolonged wet weather condition.
- Wetness for 24 hours or more and the temperature is above 36°F.
- Splashing rain are spread the disease.

Reddish Circular Spots

Whitish Gray Spots

Courtesy: **Turner Sutton**

Management

- The diseased leaves and twigs should be pruned and burnt.
- Spraying -F errous sulphate 2.5 kg + 0.5 pint sulphuric acid in 4.5 l of water.
- Spraying - BM 1.0 per cent or Copper oxychloride 0.25 per cent or Carbendazim 0.1 per cent or Mancozeb 0.2 per cent or Difolatan 0.2 per cent or Chlorothalonil 0.2 per cent or Bitertenol 0.1 per cent.
- Cultivate resistant varieties – Bangalore Blue, Beauty Seedless, Bharat Early, Golden queen, Large white.

Blue Mould Rot

Pathogen: *Penicillium digitatum*

Symptoms

- Scanty growth - white and turn bluish green are seen.
- Decay the berries.
- Infected tissues become soft and watery.

- Infected berries emit a mouldy flavor.
- The fungus covers whole berries and it looks like bluish green in colour.

Bluish Green Growth

Courtesy: **Scot Nelson**

Management

- Avoiding injuries to the ripe berries helps to reduce soft rot.
- Clean the planting stock.
- Diseased wood removal and immediate burning.
- Fungicide application at shoot extension and later if temperatures are cool.
- Sulphur is said to reduce new infections in the early spring.

Rhizopus Rot

Pathogen: *Rhizopus nigricans*

Symptoms

- Round irregular, light brown and water scaked lesion appear on fruits.
- Decaying fruits emits fermented, moudly smell.
- Conidia aseptate, small and globose.
- It is a postharvest disease.
- Under warm and moist conditions the fungus grows rapidly producing acoarse gray mat of mycelium.
- Injury caused to the berries by tight packing and storage temperature help the fungus grown during storage.
- If the infected berries are trimmed at harvest, it does not occur after harvest under ideal storage conditions.

Courtesy: **Scot Nelson**

Management

- ✰ Diphenyl sprayed on cushions is effective at 1 g and 2 g per pack in protecting the fruits upto 15 days in storage as protectant and eradicant.

Botrytis Bunch Rot or Gray Mould

Pathogen: *Botrytis cinerea*

Symptoms

- ✰ One or more berries of a cluster show signs of decay just before harvest.
- ✰ The decay may progress to include most of the berries in a cluster.
- ✰ The infected fruit may become covered with a grayish-tan powder containing the spores of the fungus.
- ✰ Berry stems and cluster stems may be invaded, causing them to shrivel.
- ✰ When the fungus decays berries low in sugar, the rotting berry has a sour odor and taste.
- ✰ If the berries are nearly mature and have a high sugar content, the decaying berry is quite firm, dry, and somewhat sweetish to the taste.
- ✰ Berries that have split or have been punctured often are attacked by other organisms, resulting in a sour or moldy decay.

Management

- ✰ Preharvest infection can be controlled by three application of Captan 0.2 per cent at monthly intervals before rainfall.

Courtesy: **Scot Nelson**

Phomopsis Cane and Leaf Spot (Dead Arm)

Pathogen: *Phomopsis viticola*

Symptoms

- Initial lesions are typically black and somewhat elliptical.
- Lesions on leaves will be circular, while lesions on petioles will be more elongate.
- Infected fruit will turn brown and become soft and shriveled.
- Small black pepper-like fungal fruiting bodies will cover the shriveled fruit.
- Symptoms of *Phomopsis* can occur on all green portions of the vine.

Courtesy: **Turner Sutton**

Black Rot

Pathogen: *Guignardia bidwellii*

Symptoms

- The disease attacks the leaves, stem, flowers and berries. All the new growth on the vine is prone to attack during the growing season.
- The symptoms are in the form of irregularly shaped reddish brown spots on the leaves and a black scab on berries.
- Occasionally, small elliptical dark coloured canker lesions occur on the young stems and tendrils.
- Leaf, cane and tendril infection can occur only when the tissue is young, but berries can be infected until almost fully-grown if an active fungicide residue is not present.
- The affected berries shrivel and become hard black mummies.

Courtesy: **Turner Sutton**

Black Mould Rot

Pathogen: *Pestalotiopsis menezesiana*

Courtesy: **TNAU Portal**

Symptoms

- ✰ The fungus enters the berries through the injuries caused due to poor postharvest handling operations.
- ✰ Berries look black in colour.
- ✰ The pulp of infected berries is reduced to watery consistency.

Green Mould Rot

Pathogen: *Penicillium canescens*

Symptoms

- ✰ It is a postharvest disease.
- ✰ The fungus enters the berries through the injuries caused due to poor postharvest handling operations.
- ✰ Infected berries are discolored yellowish green, and affected fruits shrivel.
- ✰ Masses of powdery green spores generally covered the bunch of grapes.

Courtesy: **TNAU Portal**

Soft Rots

Pathogen: *Rhizopus arrhizus* and *R. stolonifer*

Symptoms

- ✰ Small spots appear on fruits which increase in size and coalesce.
- ✰ Infection is restricted to rind, but the entire internal content decays into a pulpy mass.
- ✰ Under dry conditions, cracks originate from the point of infection.
- ✰ Packing straw should be treated with sulphur dioxide.

- Treatment of fruits with linseed oil or mustard oil or castor oil protects them from rotting.

***Courtesy*: Mike Ellis**

General Management Methods

- Sanitation is extremely important to eliminate the primary source of spores for new infections, which are found on old plant tissue.
- Proper dormant pruning and destruction of canes, clusters, and other plant parts can significantly reduce the number of spores.
- Also, canopy management during the season can help to increase airflow, which reduces free moisture and humidity within the canopy.
- Practices such as shoot positioning and strategic leaf pruning can reduce drying time.
- A complete fungicide program from pre-bloom to post-bloom is critical for managing *Phomopsis*.
- In vineyards with a history of *Phomopsis*, fungicide applications should begin at 1-inch shoot growth.

Chapter 5

Postharvest Diseases of Citrus and their Management

S. Vanitha and K. Prabakar

Department of Plant Pathology, Centre for Plant Protection Studies, Tamil Nadu Agricultural University, Coimbatore – 3, Tamil Nadu

Citrus is the World's premiere fruit crop, grown in over 100 countries on six continents. It is a non-climateric fruit and it is grown majorly in dry and intermediate zones. It belong to the family Rutaceae. Citrus fruit is enjoyed around the world for its taste, nutritional value and relatively cheap price. Normally, 30 to 40 per cent of citrus fruit loss due to lack of proper postharvest handling practices. Hence, control of postharvest diseases of citrus is vital for maintaining quality and shelf-life in a market where transport from producer to consumer may take several weeks.

Anthracnose

Pathogen: *Colletotrichium gloeosporioides*

Symptoms

- ✰ Superficial leathery appearance.
- ✰ Silver/gray to dark lesions.
- ✰ Tear-staining pattern common.
- ✰ Pink tinge (spores) under humid conditions

***Courtesy*: TNAU Portal**

Management

- ☆ Dead wood should be pruned as the fungus harbours in dead branches.
- ☆ Field sprays of copper-based fungicides should be applied prior to autumn rains.
- ☆ Postharvest treatment with Benzimidazole fungicides may reduce fruit losses.

Septoria Spot

Pathogen: *Septoria depressa.*

Symptoms

- ☆ Dark brown collapsed lesions, with a purple tinge.
- ☆ Black specks develop in decayed area.

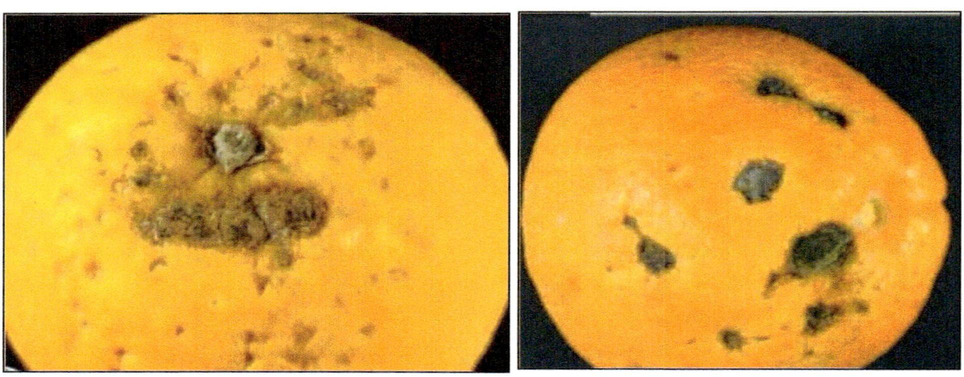

***Courtesy*: TNAU Portal**

Management

- ✰ Field application of copper-based fungicides.

Blue and Green Mould

Pathogen: *Penicillium digitatum* (green mould) and *P. italicum* (blue mould).

Symptoms

- ✰ Softening of damaged tissue.
- ✰ White fungal growth, which progressively turns blue or green as spores develop.

***Courtesy*: TNAU Portal**

Management

- ✰ Careful handling reduces damage to rind.
- ✰ Good hygiene and sorting reduces spore load and infection rates.
- ✰ Sanitation destroys spores in recirculating water and packing line equipment.
- ✰ Postharvest fungicides should be applied within 24h of harvest.
- ✰ Postharvest fungicides (Imazalil) can arrest spore development resulting in white only fungal growth.
- ✰ Lower storage temperatures slow down fungal development.

Sour Rot

Pathogen: *Galactomyctes citri-aurantii* (formally, *Geotrichum candidum*).

Symptoms

- ✰ Very soft, watery decay.
- ✰ Distinct margin between decayed and healthy tissue.
- ✰ Sour odour detectable.

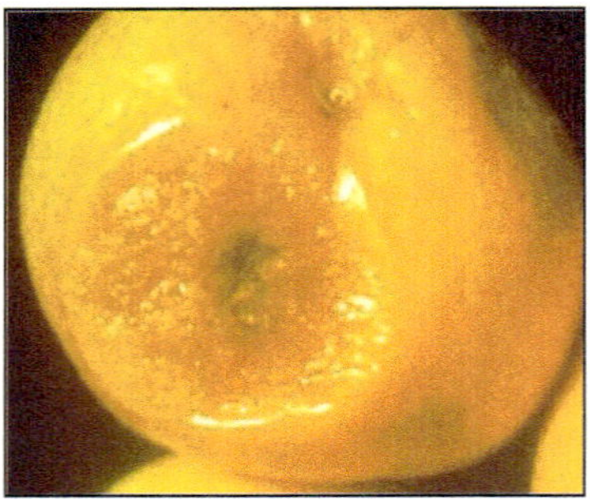
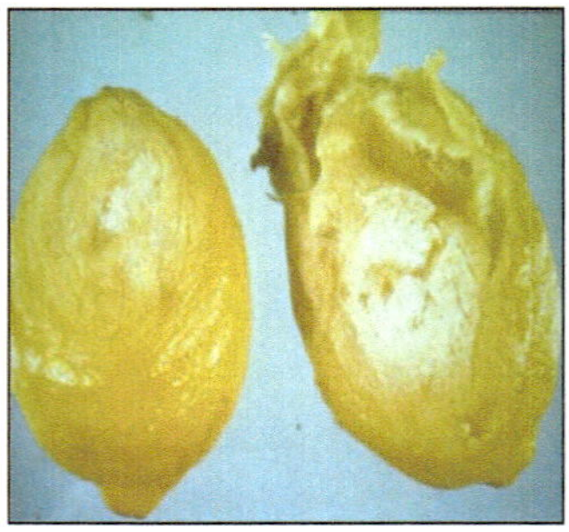
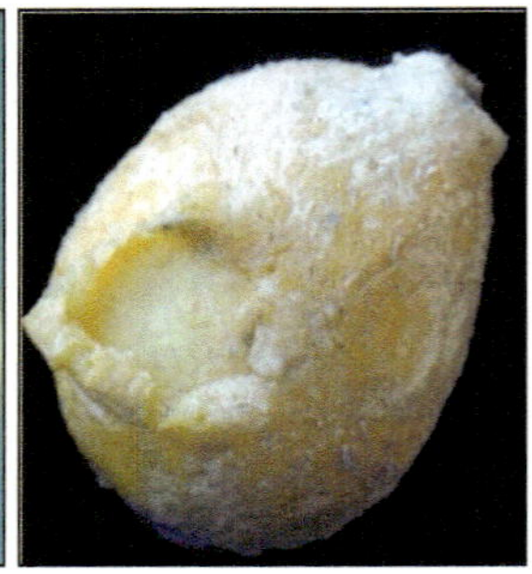

Courtesy: **TNAU Portal**

Management

- ✰ Careful handling reduces rind damage.

Stem End Rot

Pathogen: *Phomopsis* spp.

Symptoms

- ✰ A dark rot develops from the stem end as fruit ripen after harvest.
- ✰ A dark brown to black rot begins at the stem end as a dark brown ring and the rot proceeds towards the other end.
- ✰ The rot produces dark streaking of the water-conducting tissues (this symptom distinguishes stem end rot from anthracnose).

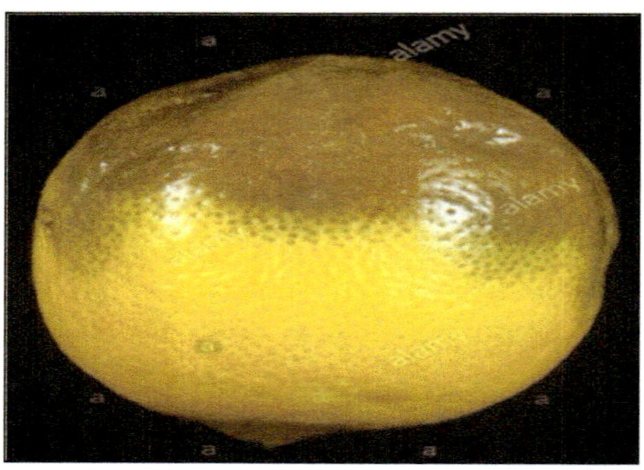

Courtesy: **G.E. Brown**

Management

- ✰ As water stress during fruit development may predispose fruit to infection, manage irrigation and root rot control carefully.

- Pre-harvest sprays of fungicides to control bacterial black spot or anthracnose may reduce the incidence of stem-end rot in fruit.
- Prune trees to improve ventilation and spray penetration.
- Remove dead branches from trees.
- Avoid harvesting immature fruit.
- Cool fruit immediately after harvest and store in well-ventilated containers.

Black Centre Rot

Pathogen: *Alternaria citri*

Symptoms

- Cracks in the stylar-end of the fruit, or the bellybutton of the fruit, allow the fungus to enter and infect the fruit.
- Infected fruit will change color prematurely and may drop early from the tree.
- A light brown to black spot on the rind can be noticed near the stylar-end of the fruit and when it is sliced open one can observe the black rot inside the fruit.

***Courtesy*: M.A. Ismail**

Management

- Preventing stress can reduce the incidence of splitting and *Alternaria* rot.
- Stylar-end infections generally occur on cultivars with poorly formed navels.
- Preharvest fungicide treatments are usually ineffective.

- Delaying harvest until infected fruit have fallen has been used as a strategy to prevent inadvertent inclusion of infected fruit in the harvested crop.
- However, unaffected fruit should be harvested at optimum maturity.
- Postharvest treatments with imazalil, 2,4-D, or both have provided some control.
- The growth regulator 2,4-D delays senescence and thereby restricts colonization of the host.

Brown Rot

Pathogen: *Phytophthora citrophthora* and *P. parasitica*

Symptoms

- The low-hanging fruit become infected first and then water will disseminate the fungus to fruit higher in the tree.
- Light brown lesions will form on the rind.
- As the disease progresses, in humid conditions, a white velvety growth can be observed on the rind and the fruit will give off a pungent, rancid odor.

Courtesy: **Don Shor**

Management

- Plant trees on a berm or high enough so that the first lateral roots are just covered with soil.
- Correcting any soil or water problems is essential for a recovery.
- Disease spread can be controlled by removing the dark, diseased bark and a buffer strip of healthy, light brown to greenish bark around the margins of the infection.

Chapter

6

Postharvest Diseases of Papaya and their Management

G. Karthikeyan and K. Prabakar

Department of Plant Pathology, Centre for Plant Protection Studies, Tamil Nadu Agricultural University, Coimbatore – 3, Tamil Nadu

Papaya (*Carica papaya* L.), a native of tropical America, is grown throughout the tropics and subtropics for its melon-like fruit. This fruit is rapidly becoming an important fruit internationally, both as fresh and processed products. Papaya fruits are very susceptible to diseases caused by many microorganisms especially fungi, as papaya fruit is high in moisture and nutrients. Orchard and postharvest diseases are very important in terms of reducing yield and quality of papaya, which are primarily responsible for the losses that occur during shipment of the fruits. Papaya postharvest losses of 10-40 per cent in sea shipments and of 5-30 per cent in air shipments are not unusual and losses due to diseases ranged from 1 to 93 per cent, depending on postharvest handling and packing procedure. Many postharvest diseases are initiated through injuries created during and following harvest. The infection process, particularly during postharvest, is greatly aided by mechanical injuries to the skin of the produce such as fingernail scratches and abrasions, insect punctures and cut. In some cases, as for latent infections, inoculation occurs prior to harvest but the disease does not develop until the postharvest period. Fruits may be infected by direct penetration of certain fungi through intact cuticle or through wounds and/or natural opening in their surfaces. For some pathogens, the synthesis of enzymes is essential for initial invasion. Furthermore, the development of fungal infection during the postharvest phase can depend upon the physiological age of the fruit, mechanical injuries, temperature and storage environment.This chapter communicating about major postharvest diseases of papaya and their management practices.

Anthracnose

Pathogen: *Gloeosporium papaya*

Symptoms

- The spots on fruits first appear as brown superficial discolouration of the skin which develop into circular, slightly sunken areas and 1 to 3 cm in dia.
- Gradually the lesions coalesce and sparse mycelial growth appears on the margins of the spots.
- Under humid conditions, encrustations of salmon pink spores are released.
- Infection at early stages of fruit results in mummification and deformation.

Courtesy: **Don Edwards**

Management

- Foliar spraying with Carbendazim 0.1 per cent at 45 days interval or
- Chlorothalonil 0.2 per cent at 10 to 15 days interval or
- Thiophanate-methyl 0.1 per cent or Mancozeb 0.2 per cent at 10 days interval controls the disease.
- Fumigation with Benzyl isothiocyanate controls postharvest spots and rots.

Rhizopus Rot

Pathogen: *Rhizopus stolonifer*

Symptoms

- ☆ Irregular, water-soaked lesions which gradually enlarge and get covered by white and dark brown fungal growth and Sporangiophores.
- ☆ The fruits become watery and emit a foul odour.
- ☆ Papaya orchard and storage room are sources of inoculum.
- ☆ Invasion occurs through wounds.
- ☆ The rots rapidly destroy the entire fruit and quickly spread to other fruits.
- ☆ Fruit fly punctures after harvest increase the disease incidence.

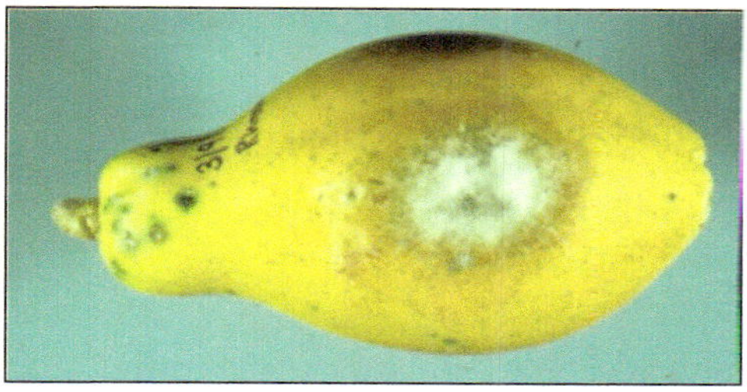

***Courtesy*: TNAU Portal**

Management

- ☆ The disease can be checked by sorting the fruit at 100°C or below but the rot appears after 2-3 days when such fruits are brought back to 200°C or above.
- ☆ Hot water treatment of fruits at 49°C for 20 minutes can check the rot.

Fruit Rot

Pathogen: *Alternaria alternata*

***Courtesy*: Don Edwards**

Symptoms

- ✰ It is characterized by gray-brown, circular to semi-circular patches covered with mycelium and conidiophores.

Management

- ✰ Foliar spray with Thiophanate-methyl 0.1 per cent or Mancozeb 0.2 per cent at 10 days interval controls the disease.

Phomopsis Fruit Rot

Pathogen: *Phomopsis caricae-papayae*

Symptoms

- ✰ Affected fruit develops a water-soaked spot which increases in size.
- ✰ The whole area becomes soft and pulpy.
- ✰ The rotten area turns dark brown to black and get depressed and cracks at a later stage.
- ✰ Water-soaking is seen in the area surrounding the diseased portion.
- ✰ All the stages of fruit are infected by the pathogen.

Phomopsis fruit rot

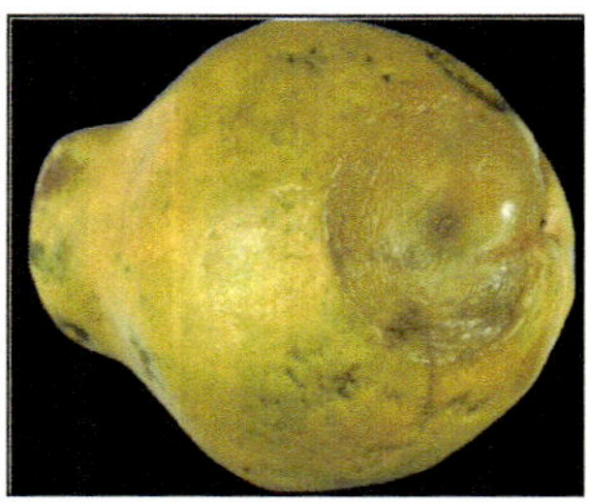

Water-soaked spot

Turns dark brown to black spots

Courtesy: **TNAU Portal**

Management

- ✰ Borax, Benlate, Captan, Difolatan and Dithane M-45 are effective.

Stem End Rot

Pathogen: *Botryodiplodia theobromae*

Symptoms

- ✰ It incites a stem end rot and a surface fruit rot.
- ✰ It usually induces a wider and softer water-soaked margin and greater internal discolouration.

- Sporulating lesions of *B.theobromae* are black and have a rough surface caused by the erumpent, confluent arrangement of pycnidia.
- The disease development is more rapid on ripe and half ripe fruits.
- The rot begins as dark green, water-soaked spots.
- Later, the affected portion becomes shriveled and turns dark brown.
- Numerous pycnidia of the pathogen appear on the diseased portion.
- The infection of fruit stalk results in fruit drop.

Stem end rot

Water-soaked margin

Black pycnidia on fruit

***Courtesy*: TNAU Portal**

Management

- Plant trees on a berm or high enough so that the first lateral roots are just covered with soil.
- Correcting any soil or water problems is essential for a recovery.
- In addition to improving the growing conditions, you can halt disease spread by removing the dark, diseased bark and a buffer strip of healthy, light brown to greenish bark around the margins of the infection.
- Hot water treatment 49°C for 20 minutes.

Chapter 7

Postharvest Diseases of Guava and their Management

T. Raguchander and K. Prabakar

Department of Plant Pathology, Centre for Plant Protection Studies, Tamil Nadu Agricultural University, Coimbatore – 3, Tamil Nadu

Guava (*Psidium gujava*) is one of the richest natural source of vitamin C containing 3 to 5 times more vitamin C than oranges and 10 times more than tomatoes and is a good source of Vitamin A and Niacin. Due to their astringent properties, mature and immature fruits, leaves, root and bark are used in local medicines to treat gastro-enteritis, diarrhoea and dysentery. High concentration of pectin in guava fruit plays a significant role in the reduction of cholesterol and thereby reduces the cardiovascular disease. Guava fruit is highly perishable so it suffers from a number of fungal diseases. Postharvest fungi deteriorate the biochemical composition of fruit.

Anthracnose

Pathogen: *Gloeosporium psidii*

Symptoms

- The most characteristic symptoms appear during the rainy season as small pin-head sized spots on the unripe fruits.
- They gradually enlarge to form sunken and circular, dark brown to black spots.
- The infected area of the unripe fruits becomes harder and corky.

- Acervuli are formed on fruit stalks.
- The pathogen remains dormant for about three months in the young infected fruits.
- In moist weather, acervuli appear as black dots scattered throughout the dead parts of the twigs.

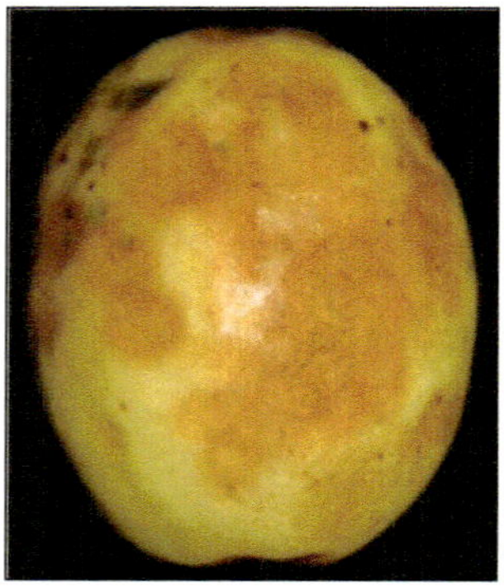

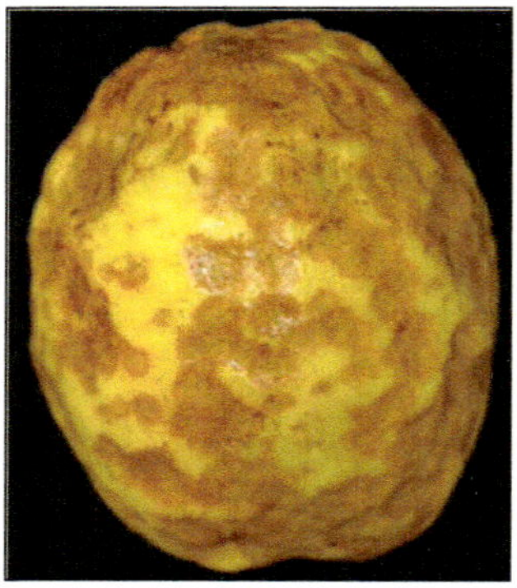

Initial symptom **Circular black spots** **Corky fruit**

***Courtesy*: TNAU Portal**

Management

- Spraying the trees with Bordeaux mixture 0.6 per cent or Copper oxychloride 0.2 per cent before the onset of monsoon reduces the disease incidence.

Phomopsis Fruit Rot

Pathogen: *Phomopsis psidii*

Symptoms

- The symptoms appear on unripe fruits at the blossom-end.
- Infected fruits show small, conspicuous, white or light brown and circular spots.
- As the fruits ripen, the spots extend and cover the fruit surface.
- The infected tissues become softer and emit an undesirable odour.

Management

- Bordeaux mixture – 0.4 per cent or Copper oxychloride - 0.2 per cent, Zineb – 0.2 per cent

***Courtesy*: N. Nongmaithem**

Botryodiplodia Rot

Pathogen: *Botryodiplodia theobromae*

Symptoms

- It causes considerable damage during storage and transit.
- The affected fruits shoe brownish discolouration mostly at the stem-end which gradually proceeds downwards in an irregular wavy-manner.
- In advanced stages, numerous, small pycnidia are produced over the entire surface of the fruit.
- The rot is soft and watery.

***Courtesy*: TNAU Portal**

Management

- Careful handling, storage at 15°C and quick disposal can minimize the losses.

Fruit Canker

Pathogens: *Glomerella psidii* and *Pestalotia psidii*

Symptoms

- ✰ Minute, brown or rust-coloured unbroken, circular, necrotic areas appear on the fruit which later tear the epidermis open in a circinate manner.
- ✰ The margin of the affected area becomes raised.

Courtesy: **K. Pernezny**

Management

- ✰ 1 per cent Bordeaux mixture or lime-sulphur (1in 25) – 3 or 4 sprayings.
- ✰ PH wash with Aureofungin (200ppm) can protect fruits for 5 days.
- ✰ The fruit rot can be checked by dip treatment of the fruits with Aretan.

Chapter

8

Postharvest Diseases of Pomegranate and their Management

A.S. Krishnamoorthy, S. Vanitha and K. Prabakar

Department of Plant Pathology, Centre for Plant Protection Studies, Tamil Nadu Agricultural University, Coimbatore – 3, Tamil Nadu

Pomegranate (*Punica granatum*) belongs to the family punicaceae bears an Aril type of fruit known as "Anar". It is one of the favourite table fruit crop of tropical and subtropical countries. Nowadays postharvest losses are increasing day by day due to one or another reasons. Postharvest losses due to fungal diseases are of prime importance which reduces the quality and quantity of marketable fruits leading to heavy losses to growers, sellers as well as consumers.This chapter epitomizing about major postharvest diseases of pomegranate and their management measures.

Alternaria Fruit Rot and Aspergillus Fruit Rot

Pathogen: *Alternaria alternata* and *Aspergillus niger*

Symptoms

- *Alternaria* fruit rot (also known as black heart) and *Aspergillus* fruit rot appear after rains as the flowers begin to open, and infect the internal portion of the pomegranate.
- Wounds must occur after flower initiation for these two infections to occur.
- The only exterior signs of disease would be a slight off-color in the peel, and a lighter weight due to internal decay.

- The hard leathery rind of infected fruit appears healthy and remains firm while the inner core of the affected fruit is partially or completely decayed starting from its calyx end.
- Affected fruit can reveal a slightly abnormal skin colour in the peel, and are of lighter weight from the internal decay.
- Heart rot infected generally float in and can at times be distinguished for the healthy fruit.

Alternaria **and** ***Apergillus*** **on Pomegranate**

Black Heart

Courtesy**: D. Edwards**

Management

- Use good orchard management practices, such as dust control and sanitation (removal of old fruit and dead branches), to reduce the incidence of the disease.

- Consider gently shaking the tree at the time of harvest to drop infected, healthy-appearing fruit to the ground. Control insect pests such as filbert worm, leaf footed bugs, and carob moth that feed on the fruit, enabling pathogen entry.
- Avoid water stress and overwatering that may result in fruit cracking.
- Avoid packing diseased fruit by thoroughly sorting and grading pomegranates for discoloration and cracking.

Gray Mould

Pathogen: *Botrytis cinerea*

Symptoms

- Gray fungal matter can be spotted on the crown area, with development of light brown or tan lesions, spreading gradually downwards from the neck of the fruit.
- The lesions become darker with time, most probably due to bleeding of the red juice from the arils as the fungus advances.
- Cracks become noticeable, as the tissue and skin expands.
- This is followed by the appearance of gray mycelial growth at advanced stages of fruit decay.
- 'Nests' can occur between adjacent fruits even in cold storage because the pathogen can grow (though slowly), at temperatures as low as –0.5 °C

Courtesy: **D. Rongai**

Management

- Use good orchard management practices, such as dust control and sanitation (removal of old fruit and dead branches), to reduce the incidence of the disease.
- Avoid water stress and overwatering that may result in fruit cracking.

- Avoid packing diseased fruit by thoroughly sorting and grading pomegranates for discoloration and cracking.
- If a postharvest fungicide is used, it can get into the crown of the fruit to prevent infection.
- Fludioxonil is registered for pomegranates and can be used to prevent gray mold.

Blue/Green Mould

Pathogen: *Rhizopus stolonifer*

Symptoms

- This fungus is characterized by powdery blue/green mould symptoms, developing from emergent circular lesions of soft, brown tissues.
- Initial symptoms include water-soaked lesions on the outer fruit surface.
- Later, a green to blue green powdery mould may appear on the surface of fruit lesions.
- Infected areas are tan or gray coloured when cut.
- At advanced stages, mycelium grows inside the fruit through connective tissues and infected arils may disintegrate into watery rot.
- The moulds can expand rapidly and are often covered with globose yellowish pycnidia that turn to dark greenish-brown over time.

***Courtesy*: D. Edwards**

Management

- Careful sanitation and avoidance of wounds, the disease may be kept under control.
- Water rinsing of fresh produce is a common practice prior to storage to remove surface dirt and spoilage pathogens.

- However pre-storage water treatment alone can actually facilitate microbial growth compared to the 'dry-treatment'.
- Chlorine washes help lessen the risk of gray mould and *Penicillium* spp. decays.
- Fungicides can be incorporated with fruit waxes to control postharvest disease and extend pomegranate shelf life in ambient and cold storage conditions.

Chapter 9

Postharvest Diseases of Sapota and their Management

G. Karthikeyan and K. Prabakar

Department of Plant Pathology, Centre for Plant Protection Studies, Tamil Nadu Agricultural University, Coimbatore – 3, Tamil Nadu

Sapota (*Manilkara achras* L.) is the important fruit crop of India next to mango, banana, citrus and grapes. Edible portion of the fruit is Mesocarp. It is a delicious dessert and nutritious fruit. Sapota is supposed to be medicinal also seeds as diuretic, bark as tonic, antipyretic, febrifuge and in curing biliousness and febrile attacks. Fruits can be dried and made into nutritious powder which can be used in Milk shakes and sweets. Sapota is good source of sugars, protein, fat, fiber and minerals (Ca, P, Fe). The fruit is rich in antioxidant polyphenolic compound tannin. Sapota fruit contains 12-14 per cent sugar. Rich source of vitamin –A (410IU/100g) and potassium (344mg/100g). Sapota contains antioxidant vitamins like vitamin-C (24.5 per cent of recommended daily intake per 100 g of fruit). It is affected by several pathogens after harvest especially, fungi which eventually reduces the quality of sapota fruit. Sapota-fruits are highly perishable and cannot be stored for long. Firm fruits can be stored up to eight weeks at 3-5°C with 85-90 per cent RH, and ripe-fruits can be stored at 2-3°C for six weeks. Low storage temperature can extend shelf-life by reducing oxidative metabolism and ethylene evolution.

Soft Rot

Pathogen: *Pestalotiopsis mangiferae*

Symptoms

- ☆ The diseases appear as water-soaked spots covering the entire fruit within 3 to 4 days.
- ☆ Rotted fruits become soft and dark brown and later numerous acervuli are seen in rotted zones.
- ☆ The fungal colonies are yellowish white.

Courtesy: **NIPHM Portal**

Fruit Rot

Pathogen: *Phytophthora palmivora, Pestalotiopsis versicolor, P. glandicola, P. sapotae* and *Botryodiplodia theobromae.*

Courtesy: **NIPHM Portal**

Symptoms

- Diseased fruits exhibited water-soaked lesions which become brown within 2 to 3 days.
- Subsequently the whole fruit is covered with tufts of mycelium.

Black Mould Rot

Pathogen: *Aspergillus niger*

Symptoms

- This disease occurs during hot summer on injured or over-ripe fruits.
- Infection progresses more or less radially on the surface.
- Later black conidial heads cover the fruit-surface and fruits become soft and crack.

Green Mould Rot

Pathogen: *Penicillium digitatum*

Symptoms

- Water-soaked brown areas appear at the proximal pedicel-end.
- Edible portions become soft, and juice leaks out with a characteristic fermentative odour; and whole fruit rots within 10-12 days.
- Maximum decay occurs at about 30°C.
- Symptoms were developed by *Penicillium digitatum* include a soft water-soaked area on the peel.
- Green asexual spores (conidia) form at the centre of the colony, surrounded by a broad band of white mycelium.
- The lesion spreads more rapidly.
- The fruit rapidly spoils and collapses, or in lower humidity shrinks and mummifies.

Rhizopus Soft Rot

Pathogen: *Rhizopus stolonifer* and *Rhizopus arrhizus*

Symptoms

- Symptoms were like as large water soaked areas which become soft and sunken. Gray whiskery mould with dusty black spores is grown on the fruit surface.

Integrated Disease Management

- Prune and destroy the dead twigs and fruits.

- Plant spacing and fertilizer régimes should be managed to avoid unnecessarily dense plant canopy.
- Prune old and non-productive branch which may serve as potential source of infection.
- Avoid injuries and bruises to fruits.
- Refrigerated storage at 10°C is found promising.
- Good field sanitation (maintain field free of infected dry or semi-dry twigs and mummified fruits of previous harvest which may serve as primary inoculum).
- Proper fertilization and irrigation, proper pruning to enhance air circulation within the canopy and sunlight penetration, managing weeds and wider tree spacing.
- Managing insect, mite and other foliar diseases increases tree.
- Remove and destroy all the affected fruits to reduce, the incidence.
- Cover the fruit with polythene bags when the fruits are up to 5 cm.
- Spray mancozeb 0.25 per cent.
- Fruit must be carefully handled in order to prevent bruising and wounding.
- Harvested fruits must not be exposed to direct sunlight, but should be kept under shade.
- Harvested fruits must be collected in clean and dry plastic crates.
- Defective fruits, *i.e.*, those that are diseased, mechanically damaged and not marketable must be separated out.
- Fruits must be cleaned by wiping or brushing.
- They must not be immersed in water as the corky fruit peel can retain moisture.
- Fruits must be transported during the cooler part of the day, in well ventilated plastic crates under clean and dry conditions.
- Wetting of fruit during transportation must be avoided.
- Fruit must be transported as quickly as possible to the final destination, to prevent spoilage.
- Fruit destined for export must be packed in cardboard cartons lined with shredded paper to protect against injury.
- Ambient conditions: For temporary storage under ambient conditions, fruit must be stored in a cool, dry place with adequate ventilation.
- Cold storage: The fruits can be transported and stored at 13–15°C and a relative humidity 85–90 per cent.

Chapter 10

Postharvest Diseases of Pome Fruit and their Management

K. Prabakar and S. Vanitha

Department of Plant Pathology, Centre for Plant Protection Studies, Tamil Nadu Agricultural University, Coimbatore – 3, Tamil Nadu

Pome fruits are members of the plant family Rosaceae. They are the fruits that have a "core" of several small seeds surrounded by a tough membrane. The membrane is, in turn, encased in an edible layer of flesh. Apples and pears are the best known examples of pome fruit; others that are found in South Africa include loquats and quinces.

The pome fruits as apple and pear represent very important fruit crops in Europe. Pome is an accessory fruit composed of one or more carpels surrounded by accessory tissue. High quality fruits in sufficient amounts can only be produced on healthy plants. Thus one of the most important task in crop production is an effective and sustainable management of plant diseases.

Lenticel Spot Disease (Bull's Eye Rot)

Pathogen: *Gloeosporium* spp., *Neofabraea* spp. and *Diaporthe perniciosa*

Symptoms

- Bull's-eye rot lesions are circular, flat to slightly sunken and appears light brown to dark brown with a lighter brown to tan center with concentric rings.
- Decayed tissue is firm.

- ✰ Cream-colored spore masses in the aged decayed area may appear.
- ✰ The symptoms are similar to that of bitter rot caused by *Glomerella cingulata* and side rot caused by *Phialophora malorum*.

***Courtesy*: J.L.J. Bezuidenhout**

Management

- ✰ Removing twigs and branches with cankers helps reduce inoculum level in the orchard.
- ✰ Recent efficacy trials conducted by WSU and Pace show that pre-harvest fungicide application failed to control bull's-eye rot in storage.
- ✰ Postharvest fungicides, TBZ (TBZ 500D) and pyrimethanil (Penbotec) by either drench or thermofog, show consistent results and effectively control bull's-eye rot.

Blue Mould

Pathogen

Several *Penicillium* species are responsible for causing blue mold rot, but *P. expansum* is the most prevalent species.

Symptoms

- ✰ Blue mold originates primarily from infection of wounds, such as stem punctures, bruises and mechanical damages from insects and birds on fruit.
- ✰ The decayed area appears a light tan to a dark brown. The decayed tissue is soft and watery.
- ✰ Blue or blue-green spore masses may appear on the decayed area, starting at the infection site.
- ✰ Decayed fruit has an earthy, musty odor.

- The presence of blue-green spore masses at the decayed area, and associated musty odor, are the positive diagnostic indication of blue mold.
- Without the blue spore masses, it is not easy to differentiate blue mold from mucor rot (*M. piriformis*) that may produce a sweet odor rather than earthy odor.

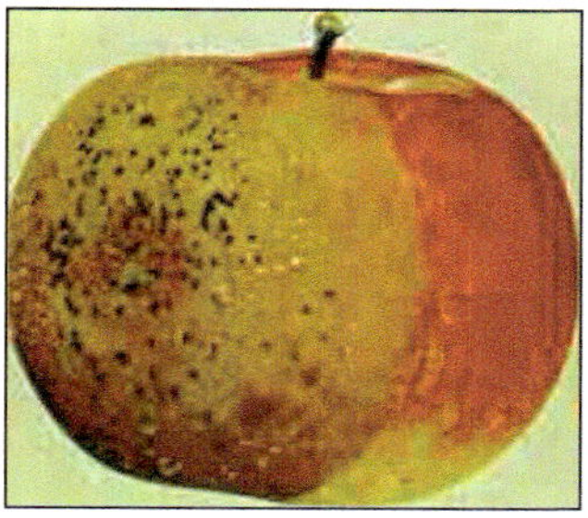

Fruit rind become watery **Fruit rot** **Affected fruits**

***Courtesy*: TNAU Portal**

Management

- Sanitizing dump-tank and flume water is an essential practice to reduce infection of fruit by *Penicillium* spp. during the packing process.
- Fruit bins and storage rooms can harbor *Penicillium* spores including fungicide resistant strains.

Gray Mold

Pathogen: *Botrytis cinerea.*

Symptoms

- The decayed area appears light brown to dark brown and the color is similar across the decayed area.
- The decayed area is firm or spongy.
- Decayed tissue is not separable from the healthy tissue.

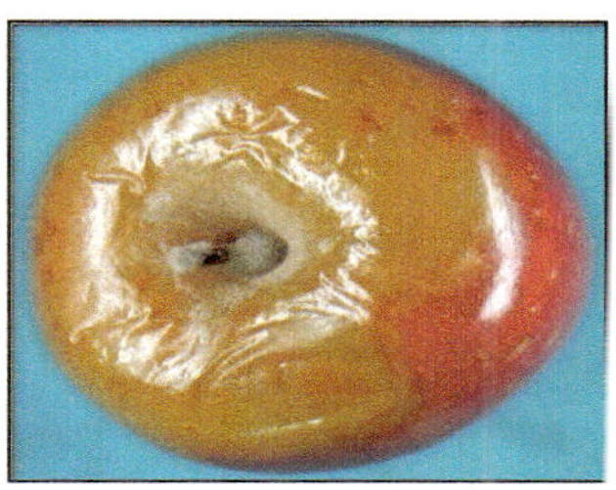
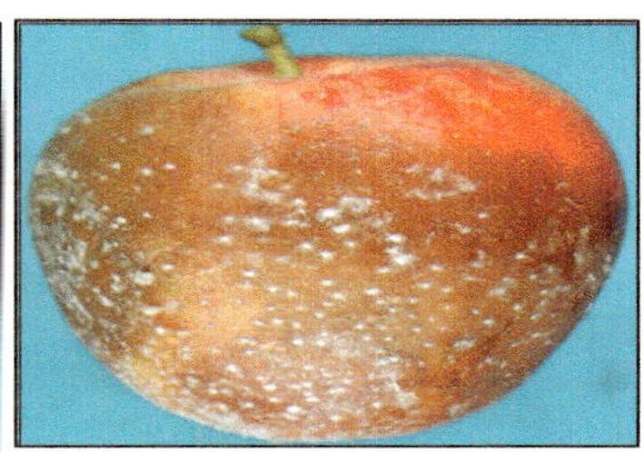

***Courtesy*: TNAU Portal**

Management

- A postharvest treatment with fludioxonil or pyrimethanil applied prior to storage, is highly effective to control gray mold, particularly for those that originate from infection of wounds.
- Preharvest spray with effective fungicides near harvest can reduce gray mold in storage.
- Orchard sanitation by removing fallen fruit and debris may help reduce the inoculum level in the orchards.

Sphaeropsis Rot

Pathogen: *Sphaeropsis pyriputrescens*

Symptoms

- The decayed area is firm or spongy with light tan to brown.
- In the advanced stage, pycnidia may form on the decayed area.
- It has a distinct odor like "used oil or bandage" from internal tissue of the decayed fruit.

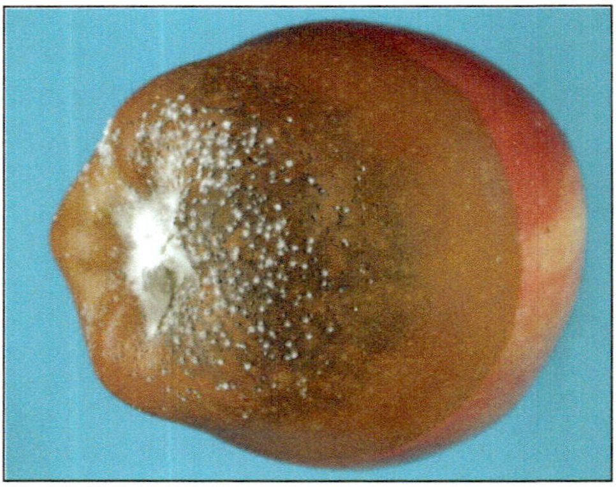

***Courtesy*: K. Peter**

Management

- Management of *Sphaeropsis* rot is based on orchard sanitation and fungicide application in pre- and/or postharvest.
- Removal of mummified fruit, twig dieback, cankers of apple and crabapple trees is very important to reduce inoculum levels in the orchard.
- Although several pre-harvest fungicides can reduce the decay incidence in storage, a postharvest fungicide application, including pyrimethanil, or fludioxonil shortly after harvest is highly recommended.

Speck Rot

Pathogen: *Phacidiopycnis washingtonensis.*

Symptoms

- The decayed tissue is firm or spongy with light tan to brown, which is very similar to gray mold and *Sphaeropsis* rot.
- Brown to black specks with a white to light tan center may present on lenticels of red apple cultivars, however gray mold apples also have similar specks.
- In the advanced stage, decayed area becomes black, full of pycnidia that are easily stuck to the surface of cardboard apple trays.
- Speck rot has a distinct odor.

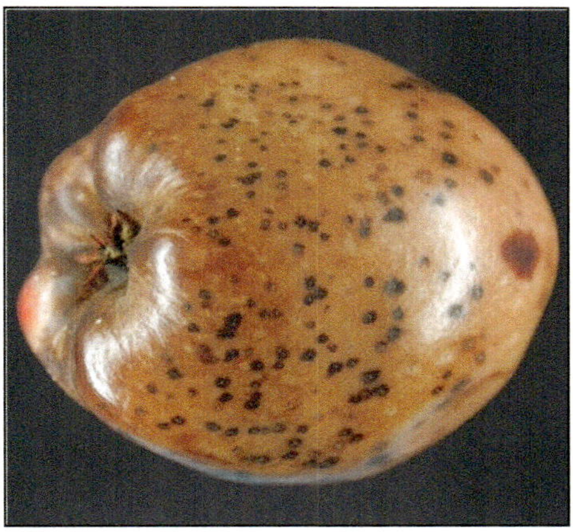

***Courtesy*: K. Peter**

Management

- Orchard sanitation is very important to reduce inoculum levels, which can be done by removing diseased plant tissues including mummified fruit, twig dieback, and cankers of apple and crabapple.
- Highly susceptible 'Manchurian' crabapples should be pruned before or right after bloom and replaced with less susceptible or resistant cultivars for a long-term management.
- Postharvest treatment with Thiabendazole, Pyrimethanil or Fludioxoil is highly effective, while preharvest fungicide application near harvest can reduce the decay incidence in storage.

Mucor Rot

Pathogen: Several species of *Mucor* are responsible for the disease, but *M. piriformis* is most common.

Symptoms

- Infected tissues of pome fruit become soft and watery.
- The decayed area appears light brown to brown with a sharp margin.
- Totaled fruit may produce alcoholic odor when kept in an airtight container.
- Gray columellate sporangia with black sporangiospores may appear on the decayed area.
- Mucor rot fruit has a sweet odor.

Courtesy: **TNAU Portal**

Management

- Orchard sanitations are important to reduce the inoculum buildup and dispersal of the fungus by removing fallen and thinned fruit.
- Fruit should not be harvested in dry weather.
- Fruit bins should not be placed on wet, muddy soil.
- Over mature fruit are highly susceptible to mucor rot and other postharvest decays.
- Fruit pickers should not place fallen or dropped fruit on the orchard ground into fruit bins and should handle picked fruit carefully to minimize stem punctures.
- Thermofogging fungicides might be the best alternative to recirculating drench to avoid cross contamination of *M. piriformis* in drench solution.
- Sanitation for dump tank and flume water helps reduce mucor rot after packing.

Green Mould Rot

Pathogen: *Penicillium italicum*

Symptoms

- It occurs on fruits during storage and transit.
- Affected of the fruits become watery.
- Watery spot increases and then entire fruit rots and emits bad smell
- Green fungal growth is seen on the surface of the fruits
- Spores of the pathogen are air-borne
- It entry through the stylar-end or lenticels was selective and in frequent.

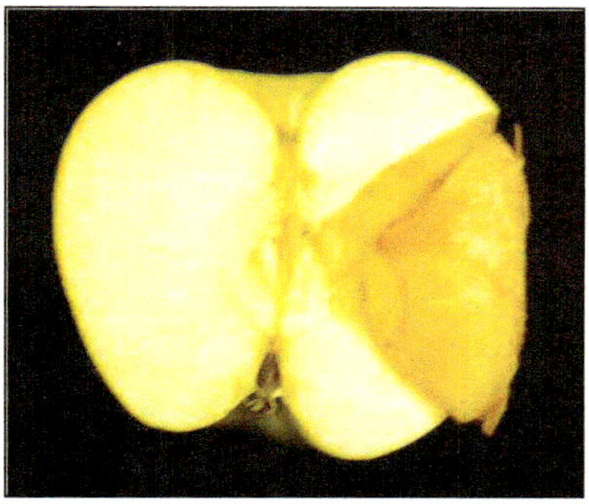

Internal watery tissue

Increased watery spots

***Courtesy*: K. Peter**

Management

- Dip the harvested fruits in 2 per cent bleaching powder – 5 minutes. Storage at 2-4°C.
- Boric acid at 4 per cent and Imazil.
- Spraying the fungicides like Benomyl 0 1 per cent or Thiophanate-methyl 0.1 per cent.

Apple Scab

Pathogen: *Venturia inaequalis*

Symptoms

- Small, rough, black, circular lesions on their skin.
- Fruits after keeping in cold storage.
- Affected fruits rot due to secondary infection of the lesions.

Black circular lesions **Enlarge spots** **Spot with Ascospores**

Courtesy: **TNAU Portal**

Management

- Infected spurs and cankers should be pruned during summer.
- Affected fruits whether on trees or fallen on the ground should be collected and buried in the soil.
- Diseased mummies should be ploughed into the soil.

Chapter

11

Postharvest Diseases of Pineapple and their Management

V.K. Parthiban, S. Vanitha and K. Prabakar

Department of Plant Pathology, Centre for Plant Protection Studies, Tamil Nadu Agricultural University, Coimbatore – 3, Tamil Nadu

Pineapple (*Ananus cosmosus*) was named for its resemblance to a pine cone. Pineapple is a well-known fruit all over the world and within the tropical fruits it represents the largest processed volume generating several kinds of products like canned and frozen pineapple.The symptoms of the postharvest diseases of pineapple fruit may develop externally or internally.

This chapter clearly implies about major postharvest diseases of pineapple and their management practices.

Fruitlet Core Rot (Green Eye)

Pathogens

- ☆ It is a postharvest fungal disease-yeast-mite complex of pineapples caused by the fungi *Penicillium funiculosum, Fusarium moniliforme* var. *subglutinas,* the round yeast *Candida guilliermondi,* the fruit mite *Steneotarsonemus,* and the red mite *Dolichotetranychus floridanus.* This disease complex is also called black spot or fruitlet brown rot.

Symptoms

- ☆ This is an internal fruit disease.

- Symptoms appear as brown to black colored areas in the center part of individual fruitlets.
- This disease is more serious in low-acid cultivars.
- Smooth Cayenne fruits do not usually show any external symptoms.
- However, fruit of the rough-leaf (Mauritius) may produce fruitlets that fail to colour – a condition often referred to as green eye.
- Severely affected fruitlets may become brown and sunken as the fruit ripens.
- Internal symptoms consist of a browning of the centre of the fruitlets starting below the floral cavity and sometimes extending to the core.
- The browning, which remains quite firm, varies in size from a speck to complete discolouration of one or more fruitlets.

Penicillium funiculosum and/or Fusarium moniliforme

***Courtesy*: Don Edwards**

Management

- The principal means of reducing the incidence of fruitlet core rot is by spraying the crop with an acaricide to control the mites which spread the disease.
- Fungicides have not been effective except when applied directly into the opening of the terminal leaves that is created by the emerging inflorescence.

Yeasty Fermentation

Pathogens: The Yeast *Saccharomyces* spp. and *Candida* spp. The yeast *Saccharomyces* is most often responsible for fermentation.

Symptoms

- Yeasts ferment sugar solution, producing alcohol and releasing carbon dioxide.
- The first symptom is a bubbling exudation of gas and juice through the crack or injury where infection occurred.

- The shell then turns brown and leathery and, as the juice escapes, the fruit becomes spongy.
- Internally, the decaying flesh turns bright yellow and develops large gas cavities.
- Finally, all that remains of the fruit is the shell and spongy tissue.

***Courtesy*: Don Edwards**

Management

- Yeasts are among the most common organisms found in nature.
- Yeasty rot is widespread but occurs mainly during spring in overripe or damaged fruit.
- Protect fruit that will ripen in spring in frost-prone areas by covering young developing fruit with paper bags.
- Fruit showing even minor interfruitlet cracking should not be consigned to the fresh-fruit market.
- Any fruit showing fractures between fruitlets should be picked at the earliest stages of fruit maturity to minimize losses.

Black Rot

Pathogen: Black rot is caused by the fungus *Ceratocystis paradoxa*. It is also known by its asexual name, *Chalara paradoxa* or *Thielaviopsis paradoxa.*

Symptoms

- It is typically the most common and severe postharvest disease of pineapple.
- It is also referred to as stem end rot, water blister, or soft rot and is characterized by a soft watery rot of the flesh.
- The severity of this disease depends on the amount of fruit bruising or wounding incurred during harvest and packing, the level of inoculum on the fruit, and the postharvest temperature.

- Black rot does not occur in the field unless the fruit is overripe or injured.
- Susceptibility also varies with cultivar, with Montserrat being less affected than Smooth Cayenne.
- Black rot usually starts at the point of detachment of the fruit (stem end), although it can readily infect bruised or other damaged areas of the fruit.

***Courtesy*: Don Edwards**

Management

- Control of black rot is achieved by minimizing mechanical injury to the fruit coupled with appropriate postharvest sanitation (150 ppm chlorine wash plus fungicide).
- The most effective postharvest fungicides for controlling black rot are thiabenbazole, benomyl, and captan.
- For effective control, the fruit must be dipped in the fungicide within 6 hours after harvest.
- Storage of the fruit at 7°C is also necessary to inhibit growth of the fungus.

Chapter

12

Postharvest Diseases of Amla and their Management

S. Vanitha, P. Muthulakshmi and K. Prabakar

Department of Plant Pathology, Centre for Plant Protection Studies, Tamil Nadu Agricultural University, Coimbatore – 3, Tamil Nadu

Amla (*Phyllanthus emblica*) is a member of Euphorbiaceae family to which most of the xerophytes belong. Amla is a hardy drought resistant fruit tree. It is also known as aonla, amali and nelli *etc.* The amla fruits are a rich source of Vitamin-C. One part or the other is used in the cure of cough, bronchitis, jaundice, diabetes, dyspepsia, diarrhea and fever. Though amla is a subtropical fruit, it thrives very well and comes to yield in tropical humid conditions also.

Amla or Indian Gooseberry is a minor subtropical deciduous tree. It thrives well throughout tropical India and is wild or cultivated in the regions extending from the foothills of Himalayas to Sri Lanka and from Malaysia to China. The tree also grows wild at elevations up to 1500 meters in southern parts on India. Amla varieties, though have not been classified according to their size, colours are named after the places of growing. The important varieties of Amla are: Banarsi, Bansi red, Chkiya, Desi, Hathi fool and Pink-tinged. The variety chakiya is noted for its heavy and regular bearing habits while banarsi is reported for its fairly large sized fruits, though it is slightly shy. Amla gets ready for harvesting by November. In the Northern plains, the peak harvesting period is from mid-December to January end. Bacterial or fungal spoilage may occur during postharvest handling of Amla fruits. This chapter rendering about the theme of major postharvest diseases of amla and their management measures.

Anthracnose

Pathogen: *Colletotrichum gloeosporioides* state of *Glomerella cingulata* (Ston.) Spauld and Schrenk.

Symptoms

- Disease appears on leaf let and fruit during August- September.
- Initial symptom of the disease is in the form of minute, circular, brown to gray spots with yellowish margin on leaflets.
- The central area of the spot remains grayish raised with dot like fruiting bodies.
- On fruits, the depressed lesions are formed, which later turn dark in the centre forming acervuli often arranged in rings.
- The lesion may vary in size and shape with spore masses appearing on fruiting bodies at high humidity.
- Consequently the fruits become shriveled and rot.
- The disease is favored by hot and humid weather.

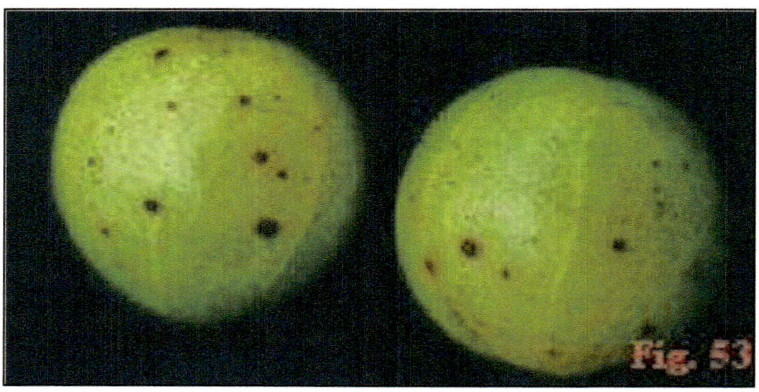

***Courtesy*: Priyamedha Sengupta**

Management

- Sanitation of orchard.
- Discard affected fruits from the orchard.
- Spray Carbendazim (0.1 per cent) before harvesting of fruits.

Soft Rot

Pathogen: *Phomopsis phyllanthi*

Symptoms

- Smoke brown to black round lesions develop on fruits within 2-3 days of infection.

- The diseased parts later show olive brown discoloration with water soaked areas extending towards both the ends of fruits forming an eye shaped appearance.
- Infected fruits become dark brown, crinkled with softening of underlining tissues and get deformed. The shape of the fruit is also deformed.
- Fungus causes infection both in young and matures fruits, but mature fruits are found to be more susceptible.
- Disease normally appears in November and December. Smoke brown to black round lesions develop within 2-3 days of infection.

***Courtesy*: Priyamedha Sengupta**

Management

- Treatment of fruits with Difolatan (0.15 per cent) or Dithane M-45 or Carbendazim (0.1 per cent) during the month of November.
- Avoid injury to fruits.

Black Soft Rot

Pathogen: *Syncephalastrum racemosum*

Symptoms

- Black Soft rot was noticed on harvested and stored (20 ± 50 C and 65 ± 5 per cent relative humidity) fruits.
- These fruits had numerous, minute brown necrotic lesions showing white mycelial growth.
- A pronounced halo of water-soaked, faded tissue surrounded the lesion between the fringe of mycelium and healthy tissue.
- The rotted surface was covered with a black, powdery layer of spores.
- Pre injured fruits are prone to infection.

Courtesy: **Priyamedha Sengupta**

Management

- Avoid fruit injury, during harvesting.
- Discard infected fruits from the orchard.
- Spray Dithane M-45 (0.2 per cent) or Carbendazim (0.1 per cent) before harvest.

Blue Mould

Pathogen: *Penicillium citrinum*

Symptoms

- It causes brown patches and water-soaked areas on the fruit surface.
- As the disease progresses, three different types of colours develop in succession, *i.e.*, first bright yellow, then purple-brown and finally bluish green.

Courtesy: **Priyamedha Sengupta**

- ☆ There is exudation of drops of yellowish liquid on the fruit surface.
- ☆ The fruits emit a bad odour.
- ☆ The entire fruit finally gives a bluish-green postulated or beaded appearance.

Management

- ☆ Careful handling of fruits. Any injury on the fruit surface during harvesting and storage make the aonla fruits prone to blue mould
- ☆ Avoid bruising or injury to fruits while harvesting.
- ☆ Sanitary conditions in storage should be maintained.
- ☆ Treatment of fruits with borax or sodium chloride (1 per cent) checks the blue mould infection.
- ☆ Treatment with carbendazim or thiophanate methyl 0.1 per cent after harvest.
- ☆ Fruits smeared with mentha oil, checks the rot.

Fruit Rot

Pathogen: *Aspergillus niger*

Courtesy: **Priyamedha Sengupta**

Symptoms

- ☆ The rot starts as a small brownish spherical necrotic spot which increases in circular fashion with the development of the disease.
- ☆ In advance stage, the spots become dark brown to black and neighboring spots coalesce.
- ☆ The central portions of the infected tissues become soft and pulpy.

Management

- ☆ One spraying of Carbendazim (0.1 per cent) should be done 15 days prior to fruit harvest.
- ☆ Harvesting should be done very carefully to avoid any injury to the fruits.
- ☆ Fruits should be stored in clean containers.
- ☆ Full sanitary measures should be adopted during storage and transit.
- ☆ Sanitary conditions in storage should be maintained.
- ☆ Treatment of fruits with borax (0.5 per cent) or sodium chloride (1 per cent) checks the fruit rots.

Postharvest Fungal Diseases of Indian Gooseberry

(I) Anthracnose by *Colletotrichum gloeosporioides*. (II) Fruit rot by *Penicillium digitatum*, (III) Soft rot by *Phomopsis phyllanthus*. (IV) Fruit rot *by Aspergillus niger.*

Integrated Disease Management

A. *At the Time of Harvest*

- ☆ Fruits should be graded according to colour, size and maturity.
- ☆ Immature fruits should be removed and set aside to ripen as some fruit species will ripen after harvest while others will tend to rot.
- ☆ Diseased and damaged fruits should be removed as these will infect the other healthy fruits.
- ☆ Fruits should be handled gently to avoid bruising even during transport.
- ☆ They should be washed in clean water and drained to remove all water. Chlorinated water (up to 100 ppm) can be used if preferred.
- ☆ To increase the self-life of the harvested fruits they can be stored at the ambient temperature in low energy cool chambers.
- ☆ Some fruits may spoil if the storage temperature is too low.
- ☆ The skin and flesh go brown due to injury through chilling.

B. Postharvest Storage

Postharvest Operations

- Remove immature, damaged and diseased fruits and grading should be done for the remaining fruits according to their size, weight, colour and maturity into three grades.
- Washing of fruits should be done using chlorinated water (100 ppm) and drain them.
- For short distance transport gunny bags or baskets (40-45 kg capacity) made of pigeon pea stems lined with newspaper as cushioning material can be used.
- For long distance transport corrugated fiber board boxes give better protection from fruit damage.

Storing of Fresh Fruits

- Depending up on the variety, the mature fruits can be stored at room temperature for about 6-9 days.
- The shelf-life can be extended up to 12-18 days if fruit are stored in low energy cool chambers.
- As mature fruits can be stored up to 2 months at 5-7°C.
- The shelf-life can be extended for up to 75 days if fruits are kept in 15 per cent brine solution (150 g salt/l) at room temperature.

Chapter

13

Postharvest Diseases of Jackfruit and their Management

K. Prabakar and S. Vanitha

Department of Plant Pathology, Centre for Plant Protection Studies, Tamil Nadu Agricultural University, Coimbatore – 3, Tamil Nadu

The jackfruit (*Artocarpus heterophyllus*) belongs to the family Moraceae also known as jack tree or simply jack. The jackfruit has played a significant role in Indian agriculture for centuries. The flesh of the jackfruit is starchy and fibrous and is a source of dietary fiber. There are two types (i.e) soft flesh and firm flesh. Fruits are harvested with a portion of the stalk attached to be used in handling them. Jackfruit fruitlets are commonly sold in producing countries as a fresh-cut product. This chapter clearly implies about major postharvest diseases of jackfruit and their management methods.

Fruit Rot

Pathogen: *Rhizopus atrocarpi, Colletotrichum lagenarium,* and *Sclerotinia rolfsii.*

Symptoms

- ✰ The problem can first be detected when a slightly discoloured soft area appears.
- ✰ Typically these soft spots are somewhat circular or elliptical in shape, and a few inches in diameter.
- ✰ In some cases depressions are visible in the skin.

- Sometime later, fungal mycelia become visible, and from there the disease progresses outward until the entire jackfruit is destroyed.
- At the early stages of the infection, the affected portion can be cut out allowing salvage of part of the fruit.
- The portion of fruit affected by rot increases naturally with the storage time.

***Courtesy*: NIPHM Portal**

Phytophthora Rot

Pathogen: *Phytopthora palmivora*

***Courtesy*: NIPHM Portal**

Symptoms

- Infection takes place through bruised or wounded skin in rough skinned varieties and through wounds in smooth skinned verities.
- Water socked lesions appear within 48-78 hours of inoculation.
- These become enlarged to form light brown spots with sporulating hyphae near the edge.
- Infected fruits develop soft rot.
- The damage caused to the bark of crown roots or bark of the trunk is called as *Phytophthora* gummosis, collar rot or foot rot.

Integrated Disease Management

- Ideally, management of jackfruit rot needs to start in the orchard.
- Most jackfruit are bagged in order to reduce attack from insects, but ideally these bags should be porous and should not allow moisture to build up next to the fruit as such conditions encourage the development of the fungal inoculums.
- Paper, cloth, or plastic- woven bags seem preferable over impermeable plain-plastic materials.
- At harvest time, the fruit must not be allowed to make contact with the soil.
- After cutting they should be placed on a clean bag or tarp, or put directly into the truck for transport to the packing house.
- The fruit should be subsequently washed in order to remove any adhering fungal spores, and then carefully dried before packing.
- Postharvest dipping in fungicide solutions such as benomyl, and simply washing in plain or chlorinated water.
- Shipping boxes should also be designed to absorb and minimize the moisture and heat build-up.
- Careful selection of ripening stages before shipping.
- Proper postharvest handling is needed, including better ways use dips and hot water washes.
- Spraying with Benomyl @ 1g/lit. completely controls the disease.

Chapter 14

Postharvest Diseases of Avocado and their Management

K. Prabakar and S. Vanitha

Department of Plant Pathology, Centre for Plant Protection Studies, Tamil Nadu Agricultural University, Coimbatore – 3, Tamil Nadu

Avocados (*Persea gratissima*) were widely in tropical America. It is quoted by the Guinness book of records as being "the most nutritious fruit known to man". The fruit is called butter pear or alligator pear and is classified as a berry. It is a climacteric fruit which matures on the tree but ripens off the tree. A typical serving of avocado (100g) is moderate to rich in several B vitamins,vitamin K and potassium.

Anthracnose

Pathogen: *Colletotrichum gloeosporioides* and *C. acutatum* (Telomorph: *Glomerella cingulata*)

Symptoms

- ✰ It is difficult to distinguish visually between the two species at the young lesion stage, but on well-developed rots the growth of *C. gloeosporioides* is more diffusing, and the orange to pink coloration duller than that of *C. acutatum*.
- ✰ Avocados will suddenly sport small dark spots that expand rapidly, within just one or two days, as the fruit ripen.

- Because the skin of the immature avocado fruit is largely protective against anthracnose infection, it's easy to have a bad case of anthracnose without even knowing it.

***Courtesy*: Tom A. Zitter**

Management

- Remove all dead fruits, leaves and branches at the end of the year and cleaning up any debris or dropped fruits.
- Prune your trees so the insides are more open and allow the wind to penetrate, reducing life-giving humidity in the canopy.
- Spraying the tree with copper fungicide every two weeks after blossom drop will ensure that your fruit is protected throughout its development.
- Also, treating or remedying other diseases, pests or handling problems will also help tremendously.
- Fruits should be carefully handled after harvest.
- Cooling ripening fruits immediately and holding them at 5°C.

Bacterial Soft Rot

Pathogen: *Erwinia carotovora* var. *aroideae.*

Symptoms

- Gray to black, mushy, foul smelling rot on fruit; fruit has darkened metallic sheen.
- It produces lesions similar to those of the original avocado fruit.

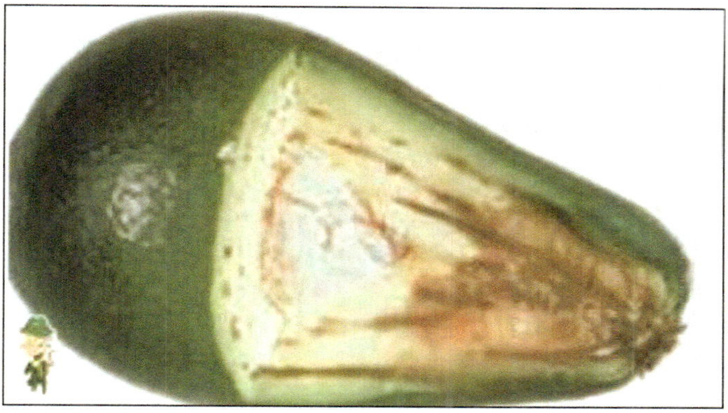

***Courtesy*: Dr. Mary Lu Arpaia**

Management

- ✰ Use certified seed, disinfected tools and cuttings.
- ✰ Employ crop rotation.
- ✰ Remove plant debris from soil.

Stem-End Rot

Pathogen

- ✰ Many different fungal species: *Botryosphaeria dothidea, Dothiorella* spp., *Glomerella cingulata, Lasiodiplodia theobromae, Stilbella cinnabarina, Thyronectaria pseudotrichia, Phomopsis perseae.*

***Courtesy*: J.W. Shepherd**

Symptoms

- ✰ Shrivelled tissue at stem end; dark brown or black lesions at stem end and eventually over entire fruit; fruit covered in mycelium.

Management

- ✰ Prune dead limbs and twigs; prune and harvest in dry conditions; provide trees with sufficient irrigation; apply a thick layer of mulch.

Chapter

15

Postharvest Diseases of Stone Fruits and their Management

K. Prabakar and S. Vanitha

Department of Plant Pathology, Centre for Plant Protection Studies, Tamil Nadu Agricultural University, Coimbatore – 3, Tamil Nadu

Stone fruits are highly perishable and decay is the major problem in postharvest life extension of fruits for the fresh market and for the processing procedure. Postharvest quality in stone fruits is ultimately defined in terms of consumer satisfaction, and includes appearance, texture and flavor, and nutritional value and safety. These result from all decisions and practices beginning with site and cultivar selection, and extending through cultural practices and all stages of postharvest handling. The development of high quality fruit is influenced by such cultural practices as mineral nutrition, irrigation, tree training, pruning and fruit thinning. Harvest maturity is of special importance because of the detrimental effects on quality of harvesting at either too low or too high a maturity, and low maturity harvest is not a solution to long distance marketing. Fruit deterioration, from such factors as fruit rotting organisms, water loss, mechanical bruising and physiological disorders, will adversely impact quality in the market. Of special concern is the need to rapidly cool stone fruits and maintain them near 0°C throughout marketing. Only by attention to all aspects of production and postharvest handling can high quality be achieved and protected in fresh stone fruits. People's perception of quality will change as the fruit pass through various stages of production, distribution and consumption, and will be more and more affected by the distribution system itself. Appearance, flavor and consumer satisfaction will remain important, but with prolonged handling they may be increasingly influenced by results of abuse: bruising, shrivel, internal breakdown, fruit rotting, *etc.*

Brown Rot

Crop: Peach, Cherry fruit

Pathogen: *Monilia fructicola*

Symptoms

- ✰ Brown rot sometimes occurs as quiescent infections which can be detected as small, circular, necrotic lesions on immature fruits.
- ✰ Immature or mature fruit with brown rot infections will sporulate profusely, shrivel, and become tough grayish-black mummies.
- ✰ Decaying fruit in cold storage or transit may appear black with little or no sporulation.

***Courtesy*: D.F. Ritchie**

- ✰ Brown rot producing spores on the surface of an infected peach fruit.
- ✰ These spores can infect other fruit.

Brown rot on cherry fruit

Brown rot mummy

***Courtesy*: K. Peter**

Management

- Sanitation is essential.
- Remove all remaining fruit from the tree after the final picking.
- This practice limits infection of fruit peduncles and twigs thus reducing the number of brown rot cankers.
- Removal of remaining fruit after final picking separates the practice of removing mummies from spring pruning.
- Avoid dumping rotten fruit in one location, which could become the starting point for disease and insect outbreaks in the following season.
- Pick and handle fruit carefully to avoid injuries; remove field heat from the fruit promptly after harvest by hydrocooling or forced air cooling; use clean containers; keep packing areas clean.
- Fruit should be cooled and refrigerated immediately after harvest.

Gray Mold

Crop: Strawberry

Pathogen: *Boyrytis cinerea*

Symptoms

- Disease symptoms appear as grayish colored soft, mushy spots on leaves, stems, flowers and on produce.
- Spots may become covered with a coating of gray fungus spores, especially if humidity is high.
- Fruit or plants shrivel and rot and often develop black, stone-like sclerotia - a compact mass of hardened fungal filaments - under rotted parts.

- The disease may also occur in storage areas causing rotting of harvested fruits and vegetables.

Courtesy: **Steven Koike**

Management

- Prune or stake plants to improve air circulation between plants. Make sure to disinfect your pruning equipment (one part bleach to 4 parts water) after each cut.
- If growing indoors, use a small clip-on fan to improve air flow.
- Keep the soil under plants clean and rake up any fallen debris.
- Add a good amount of organic compost or mulch under plants. Mulches will prevent the fungal spores from splashing back up onto flowers and leaves.
- Water in the early morning hours, or use a soaker hose, to give plants time to dry out during the day.
- Do not compost infected plant leaves or stems, and thoroughly clean up garden areas in the fall to reduce over wintering sites for the fungal spores.
- Copper-Soap fungicides will help by protecting plants from disease spores.
- Mycostop, a biological fungicide approved for organic use, has shown suppression of the disease. Direct spray (5 gm/50 liters of water) to susceptible leaves, flowers and fruits to point of run-off. Apply every 2-3 weeks, or as needed depending on disease pressure.

Rhizopus Rot

Crop: Cherry

Pathogen: *Rhizopus stolonifer*

Symptoms

- *Rhizopus* rot begins much like brown rot—as a small, brown, circular spot—but with a detectable difference.
- The skin of *Rhizopus* rot-infected fruit slips readily from the underlying flesh, while the skin of brown rotted areas is tough and leathery.
- At normal temperatures, the small spots of *Rhizopus* rot enlarge rapidly and can involve the entire fruit in 24 to 48 hours.
- A white, whiskery mold appears on the surface of infected fruits, spreading to nearby fruit and the walls of the container.
- By this time the fruit tends to leak and to smell like vinegar.
- Finally, tiny, black, spherical structures are produced on stalks above the white mold.
- Each of these contains thousands of spores (sporangiospores) that are released to float in the air. At this stage the mold looks mostly black.

***Courtesy*: K. Peter**

- Tiny, black, spherical structures are produced on stalks above the white mold.

Management

- More importantly, preventing skin cuts and punctures during harvest and packing is prime in controlling *Rhizopus* rot.
- Clean containers and good house-keeping in the packing shed and storage will aid greatly in reducing the spore population.
- Quickly removing field heat drastically slows decay, as does refrigerating fruit until it is sold to the consumer or is processed.

Blue Mould

Crop: Peach

Pathogen: *Penicillium expansum*

Symptoms

- It is characterized by a light brown discoloration of the skin followed by a soft and wet rot that can rapidly affect the entire berry.
- On the surface of rotted berries, blue-green mold colonies, with or without synnemata, may appear.
- An internal watery breakdown often occurs in grapes stored for several weeks at 0°C.
- Mycelial growth of *P. expansum* can be considerably arrested, but not totally inhibited, at 0°C.
- Blue mold infections can be initiated in the vineyard, at the packinghouse or during storage, and they are always associated with wounded fruits.

***Courtesy*: K. Peter**

Management

- Clean containers and good housekeeping in the packing shed and storage will aid greatly in reducing the spore population.
- Water in the early morning hours, or use a soaker hose, to give plants time to dry out during the day.
- Do not compost infected plant leaves or stems, and thoroughly clean up garden areas in the fall to reduce over wintering sites for the fungal spores.
- Copper-Soap fungicides will help by protecting plants from disease spores.

Postharvest Diseases of Vegetables

Chapter 16

Postharvest Diseases of Tomato and their Management

K. Kavitha, S. Vanitha and K. Prabakar

Department of Plant Pathology, Centre for Plant Protection Studies, Tamil Nadu Agricultural University, Coimbatore – 3, Tamil Nadu

Tomato (*Solanum lycopersicum* L.) is one of the most widely cultivated and extensively consumed horticultural crops globally. By weight, tomatoes rank second only to potatoes in global production of all horticultural produce. Tomato and tomato-based foods provide a wide variety of nutrients and many health-related benefits to the body. Tomato contains higher amounts of lycopene, a type of carotenoid with antioxidant properties which is beneficial in reducing the incidence of some chronic diseases like cancer and many other cardiovascular disorders. Tomato production has increased in recent years to about 163 million tonnes by 2013 due to the economic and nutritional importance of the crop. However, scientific research has focused on production mainly whilst neglecting postharvest issues. Postharvest loss is a major challenge hampering tomatoes production in most developing countries. Tomato being a perishable crop as a result of its high moisture content has short shelf life of about 48 hours under tropical conditions. Specialized postharvest handling practices and treatment methods are needed in order to extend the shelf life of the crop after harvest. Failure to adhere to these specialized handling practices and treatment methods will result in high amount of loss. Losses of up to 50 per cent can be recorded in tomatoes between the harvesting and consumption stages of the distribution chain in tropical countries. It is therefore, important to know the appropriate handling practices and treatment methods needed for harvested tomatoes in order to reduce postharvest losses thereby increasing profitability for handlers in developing countries.

Gray Mould

Pathogen: *Botrytis cinerea*

Symptoms

- Lesion-a watery area with a light brown or tan-colored central region.
- Converted in to a soft, watery mass within a few days.
- Skin is broken, the grayish mycelium and spore clusters develop within a few hours.
- Halo forms around the point of entry-small whitish rings approximately-develop on young green fruit.
- "Ghost spots" are usually single rings but may be solid white spots, the center of which contains dark-brown specks.

Courtesy: **Abdul Hafeez**

Management

- Spraying with Bordeaux mixture 1.0 per cent or Mancozeb 0.2 per cent is helpful in reducing the disease.

Early Blight

Pathogen: *Alternaria solani*

Symptoms

- The fruit become infected –through the calyx or stem attachment, either in the green or ripe stage.
- Concentric ring present on the fruit surface.
- Appear leathery and maybe covered by a velvety mass of black spores.

- Infected fruit frequently drop, and losses of 50 per cent of the immature fruit may occur.

***Courtesy*: Abdul Hafeez**

Management

- Disease free seeds should be used for sowing.
- Seeds soaked in Thiram 0.2 per cent at 30°C for 24h give better protection.
- Seed treatment with Thiram 2g/kg gives good protection against seed borne infection.
- Three sprayings with Difolatan 0.2 per cent or Mancozeb at fortnightly interval prevent the spread of the disease.
- Infected plant debris should be removed.
- Three year rotation with non solanaceous crop is recommended.

Bacterial Soft Rot and Hollowstem

Pathogen: *Pectobacterium carotovorum* subsp. *carotovorum*

Symptoms

- Fruit-soft watery decay of fruit, starting a tone or more points, as very small spots. Enlarge-very rapidly until the entire fruit-soft watery mass.
- Pathogen liquefies fruit tissue by breaking down the pectate "glue" that holds plant cells together.
- Leakage-internal collapse resembling a shriveled water balloon.

Management

- Use of chlorinated water to reduce the risk of infection during washing.
- Harvest during dry weather and minimize fruit injury at harvest.

Courtesy: **Abdul Hafeez**

Phoma Rot

Pathogen: *Phoma destructive*

Symptoms

- Distinguished from other rots by the black color or this spot.
- Small, black, pimple-like eruptions.
- Specks are the pycnidia or fruiting bodies of the fungus.
- Moderate temperature and high humidity.

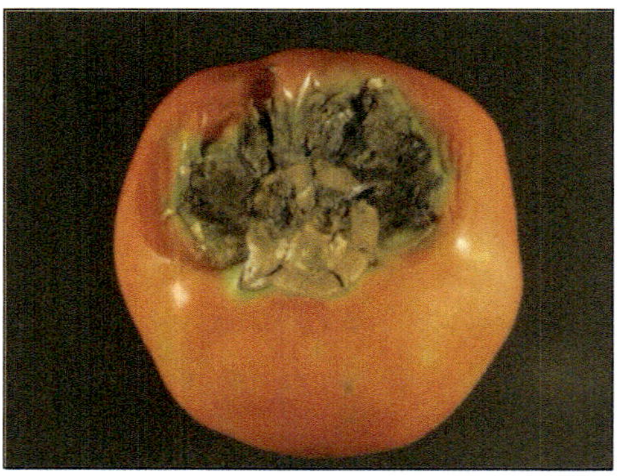

Courtesy: **Abdul Hafeez**

Management

- Seed treatment with zineb 0.2 per cent gives adequate protection against the disease.

Rhizopus Rot

Pathogen: *Rhizopus stolonifer*

Symptoms

- On tomatoes, Rhizopus rot appears water-soaked and may exude a clear liquid.
- Lesion surface may be covered with thin, cotton-like fungal structures (especially under humid conditions).
- Dark sporulation may crown the white tuft of *Rhizopus.*
- Mycelium can infect adjacent fruit through natural openings or mechanical wounds, creating nests of mold and diseased fruit.

***Courtesy*: Abdul Hafeez**

Management

- When the fruits are transported to long distance, they should be dipped in 0.2 per cent Mancozeb.
- Effective check of rots is also achieved with Chlorothalonil (0.1 per cent) treatments.
- Treatment with Thiobendazole 1000 ppm has been found to be effective in controlling the storage disease of tomato caused by *Colletotrichum dematium*.
- Spraying of Streptomycin (Agrimycin 100) @1000 ppm or Carbendazim(0.2 per cent) at 10 days interval provides good control of bacterial diseases.
- A number of bacterial and fungal antagonists have been found to effectively control postharvest rots of tomatoes.

- They have various modes of action that include antibiosis or competition for nutrients and space.
- This method of control for postharvest rots of vegetables appears to be feasible and may present an alternative to synthetic pesticides.
- Application of *Trichoderma harzianum* in the field and also as a dip before storage has been recorded to be quite effective.

Chapter 17

Postharvest Diseases of Brinjal and their Management

K. Kavitha, S. Vanitha and K. Prabakar

Department of Plant Pathology, Centre for Plant Protection Studies, Tamil Nadu Agricultural University, Coimbatore – 3, Tamil Nadu

Brinjal or Eggplant (*Solanum melongena* L.) is an important solanaceous crop of sub-tropics and tropics. Brinjal is an important vegetable crop of the Far East, Bangladesh, India, China and the Philippines. The name brinjal is popular in Indian subcontinents and is derived from Arabic and Sanskrit whereas the name eggplant has been derived from the shape of the fruit of some varieties which are white and resemble in shape to chicken eggs. Asia has the largest eggplant production which comprises more than 90 per cent of the world production area and 87 per cent of the world production. In India, it is one of the most common, popular and principal vegetable crops grown throughout the country except higher altitudes. It is a versatile crop adapted to different agro-climatic regions and can be grown throughout the year. It is a perennial but grown commercially as an annual crop. Bitterness in eggplant is due to the presence of glycoalkaloids which are of wide occurrence in plants of Solanaceae family. The glycoalkaloid contents in the Indian commercial cultivars vary from 0.37 mg/100 g fresh weight to 4.83 mg). Generally, the high content of glycoalkaloids (20 mg/100 g FW) produce a bitter taste and off flavor. The discoloration in eggplant fruit is attributed to high polyphenol oxidase activity. Brinjal fruit (unripe) is primarily consumed as cooked vegetable in various ways and dried shoots are used as fuel in rural areas. It is low in calories and fats, contains mostly water, some protein, fiber and carbohydrates. It is a good source of minerals and vitamins and is rich in total water soluble sugars, free reducing sugars, amide proteins among other nutrients. Brinjal being a perishable crop as a result of its high moisture content has short shelf life of about 48 hours under tropical conditions. Specialised postharvest handling practices and treatment methods are needed in order to extend the shelf life of the crop after harvest. Failure to adhere to these specialised handling practices and treatment methods will result in high amount of loss.This chapter clearly discussing about major postharvest diseases of brinjal and their effective management techniques.

Fruit Rot

Pathogen: *Phomopsis vexans*

Symptoms

- Affects all above the ground plant parts. Spots generally appear first on seedling, stems or leaves.
- Girdle seedling stems and kill the seedlings.
- Leaf spots are clearly defined, circular, up to about 1inch in diameter, and brown to gray with a narrow dark brown margin.
- Fruits pots are much larger, affected fruit are first soft and watery but later maybe come black and mummified.
- Center of the spot becomes gray, and black pycnidia develop.

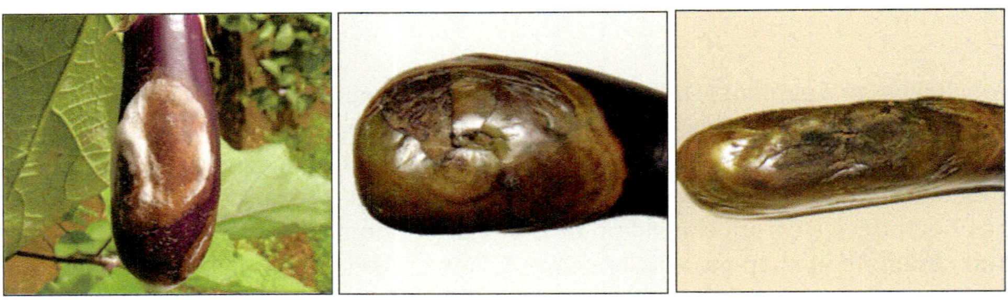

***Courtesy*: TNAU Portal**

Management

- Seeds should be dipped in hot water at 50°C for 30min. spraying with difolation 0.2 per cent or captan 0.2 per cent in the nursery and field at 7-10 days interval controls the disease.
- Deep summer ploughing, three year crop rotation and collection and destruction of diseased plant debris are some of the other control methods.
- Spraying the crop in the field with zineb 0.2 per cent or Bordeaux mixture 0.8 per cent is effective in controlling *Phomopsis* blight.

Anthracnose

Pathogen: *Colletotrichum melongenae*

Symptoms

- ✰ Sunken lesions on fruits vary in size up to 1.3cm wide and may coalesce.
- ✰ Tan colored ooze of fungal spores appear on lesions.
- ✰ Fruit dries and become black – fruit drop.

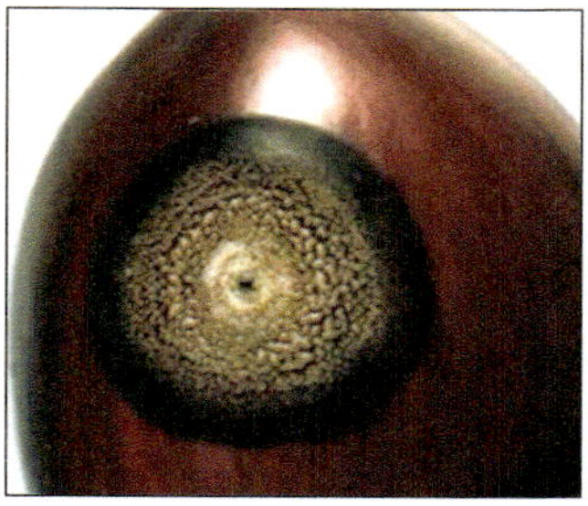

Concentric rings

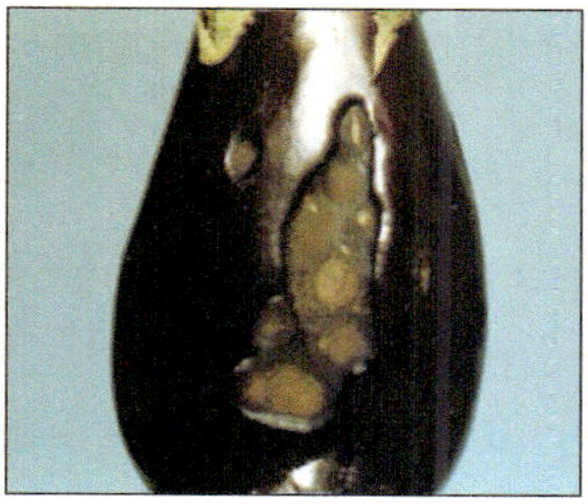

Decayed fruit

Severe infection

Courtesy: **TNAU Portal**

Management

- ✰ Do not grow egg plant or other solanaceous plants in the same area for 3 to 4 years.
- ✰ Use drip irrigation rather than over head irrigation.
- ✰ Destroy all plant debris by incorporating in the soil.
- ✰ Select disease free seeds.
- ✰ Spray with Carbendazim 1g/liter or Benomyl 1g/liter or Mancozeb 2g/liter of water when fruits reach size of a quarter at 15 days interval.

Fruit Rot

Pathogen: *Phytophthora nicotianae* var. *nicotianae*

Symptoms

- ✰ Fruits - Small, water soaked lesions.
- ✰ Skin - discoloration, turns grayish brown, and develop white cottony mycelium in humid weather.
- ✰ Rotten parts become depressed and develops wrinkles.
- ✰ Rotting spread deep into the pulp and turns brown, watery and soft.
- ✰ Rapidly spreads during rainy season.

Skin discolouration

White cottony mycelium

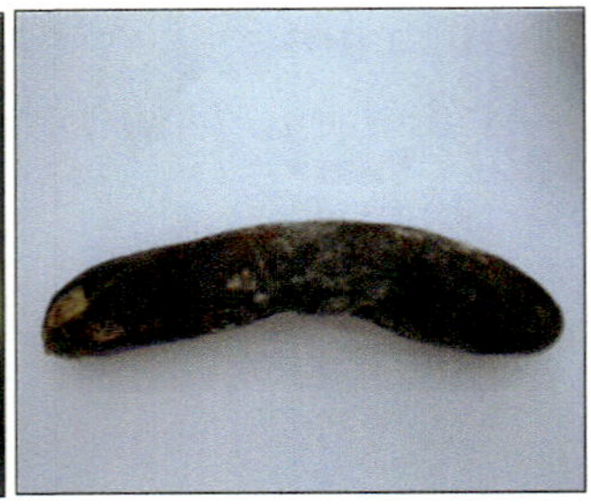

Rotten brinjal

***Courtesy*: TNAU Portal**

Management

- Removal and destruction of the affected fruits.
- Spraying the crop with Difolatan (0.3 per cent) thrice at an interval of 10 days effectively controls the disease.

Chapter

18

Postharvest Diseases of Bhendi and their Management

M. Karthikeyan and K. Prabakar

Department of Plant Pathology, Centre for Plant Protection Studies, Tamil Nadu Agricultural University, Coimbatore – 3, Tamil Nadu

Okra (*Abelmoschus esculentus* (L.) Moench) as lady's finger, is an important vegetable crop under tropical and subtropical conditions including India. Crop suffers from a number of phytopathogenic fungal and bacterial species causing severe losses, reduces plating and market value of okra seeds. The diseases of okra caused by microorganisms *i.e.* fungi, bacteria, viruses either due to contaminations and infections occur during growing season or due to infection incidental during harvesting, processing and packing or transportation process. The amount of damage and loss associated with the diseases varies greatly with commodity, processing, growing conditions and the way of handling. The postharvest diseases spooned during transit and storage throughout the country and sources of continuous drain on our already acute food situation and lead to a waste of labor, time, land and money. In India, no exact estimates of such losses are available for postharvest or storage diseases of vegetable. Since the associated microflora reduces the quality, price rate or market value of this vegetable therefore aim of present investigation is to study the major fruit diseases in storage and their postharvest impact on stored fruits.

Phoma Canker

Pathogen: *Phoma exigua*

Symptoms

- Water soaked lesion appear on fruits.
- Black spots with irregular margin.
- Black area- pycnidial formation.
- 80-90 per cent fruit loss postharvest rot of okra pods *Rhizoctonia solani* in Brazil.
- Completely rotted, the pod's typical greenish color turning brown and the infected tissues fully covered with mycelia.
- Internally, immature seeds and placenta infected.
- Diseased tissues were light brown to black.
- Externally, mycelia tend to be fluffy and lighter in color, forming a large number of dark sclerotia on the fruit surface.

***Courtesy*: Ahmed Bilal**

Alternaria Rot

Pathogen: *Alternaria alternate*

Inner syndrom of pods

***Courtesy*: Ahmed Bilal**

Symptoms

- Dark gray, slightly sunken irregular lesion, developing lengthwise.
- Gray cottony mycelial mat bearing conidia appeared on the surface of developing lesions.
- When diseased pod cut transversally – blackened necrotic tissues.

Rhizoctonia Rot

Pathogen: *Rhizoctonia solani*

Symptoms

- Pod - greenish color turning brown, and the infected tissues fully covered with mycelia.
- Internally, immature seeds and placenta were infected and the diseased tissues were light brown to black.
- Externally, mycelia tend to be fluffy and lighter in color, forming a large number of dark sclerotia on the fruit surface.
- Pods showing distinct degrees of rotting.
- Rotting of placenta tissues and immature seeds.

Brownish pod internal tissue with mycelia

***Courtesy*: Ahmed Bilal**

Management

- Plant debris from previous crops should be removed before planting.
- Seed treatment with Thiram or Captan 2g/kg of seeds.
- Use of resistant varieties.
- Field sanitation.

Chapter

19

Postharvest Diseases of Chilli and their Management

K. Kavitha and S. Vanitha

Department of Plant Pathology, Centre for Plant Protection Studies, Tamil Nadu Agricultural University, Coimbatore – 3, Tamil Nadu

Chilli (*Capsicum frutescens*) is an important spice crop and India is one of the leading producer and exporter of chilli in the world. Chilli is widely used around the world in food as a spice both in fresh and dried form which adds flavour to the meal by creating spicy pungent taste. The chemical components of the chilli may vary considerably depending on the location of cultivation and postharvest treatments. Chilli contains polyphenol compounds such as capsaisin capsorubin, zeaxanthin and cryptoxanthin are responsible for color in Capsicum species that accounts for its characteristic aroma and therapeutic properties. Postharvest losses in chilli are estimated to be 25-35 per cent. Due to postharvest losses farmers lose both in quality and quantity of the chilli. This compromises farmer's ability to market their produce.

Bacterial Soft Rot

Pathogen: *Pectobacterium carotovorum* subsp. *carotovorum*

Symptoms

- ✰ The fleshy fruit peduncle is highly susceptible and is frequently the initial point of infection.

- Both ripe and green fruit may be affected.
- Initially, the lesions on the fruit are light to dark-colored, water-soaked, and somewhat sunken.
- The affected areas expand very rapidly, particularly under high temperatures, and tissues lose their texture.
- In later stages, bacterial ooze may develop from affected areas, and secondary organisms follow, often invading the rotted tissue.
- Postharvest softening of stem end of fruit.
- The affected fruit hang from the plant like a water-filled bag.

Courtesy: **Marita Cantwell**

Management

- Use chlorinated wash water to reduce populations of soft rot bacteria and to reduce the risk of infection during washing.
- This will not reduce soft rot development in fruit infected with the bacterium prior to harvest.
- Allow fruit to dry thoroughly.
- During packing and storage, the fruit should be kept clean and maintained in a cool, dry place.
- Seed treatment – 1 per cent sodium hypochlorite for 30 sec, then rinse with clean water
- Avoid planting pepper crops following crops of potato or cabbage.
- Rotate instead with crops of bean, corn and soybean.

Alternaria Rot

Pathogen: *Alternaria solani*

Symptoms

- The fungus is reported to enter wounds (sun scald or punctures). Dusty black spores on fruits pots are characteristic.
- In most instances this disease follows blossom- end rot, but it also follows injuries, chilling, and other decays.
- On the fruit, large greenish –brown to brown lesions covered, with grayish-brown mold are produced.
- Similar lesions on the lower-part of the fruit are characteristic of *Alternaria* rot following blossom-end rot.
- The larger lesions may show alternating light and dark-brown concentric zones.
- Shipping peppers under standard refrigeration will check the development of this rot, but when the fruit is removed from refrigeration the decay will advance rapidly at moderate to warm temperatures.

***Courtesy*: Marita Cantwell**

Management

- The effectiveness of a pre storage dry heat treatment and hot water dip in reducing storage rots of capsicum caused by *Alternaria alternata*.
- Treatment with hot air at 38°C for 48-72h or hot water at 50°C to 53°C for 2 to 3 minutes, resulted in reduction in the pathogenicity and development of these pathogens in inoculate peppers.

- ✰ Fortnightly spraying of Bordeaux mixture 1.0 per cent or Copper oxy chloride 0.3 per cent or Difolatan 0.3 per cent or Mancozeb 0.2 per cent.

Anthracnose

Pathogen: *Colletotrichum capsici*

Symptoms

- ✰ Ripe fruits turning red are affected.
- ✰ Small, black, circular spot appears on the fruit skin.
- ✰ Badly diseased fruits turn straw colour or pale white colour, lose their pungency.
- ✰ Diseased cut open fruits - lower surface of the skin is covered with minute, elevated sclerotia observed.
- ✰ Advanced stage - seeds covered by a mat of fungal hyphae, turn rusty.

***Courtesy*: Marita Cantwell**

Management

- ✰ Use disease free seeds.
- ✰ Seed treatment - Thiram 2 kg/ha or Zineb 2.5 kg/ha.
- ✰ Three sprayings with captan 0.2 per cent 1st spraying - just before flowering, 2nd at the time of fruit formation, 3rd - fortnight interval after second spraying.

Gray Mould

Pathogen: *Botrytis cinerea*

Symptoms

- ✰ Brownish spots develop near the soil line or cotyledons.

- Water-soaked lesions on leaves and stems darken and collapse.
- Water-soaked spots that rapidly expand into large yellowish-green or grayish-brown, irregular lesions that are soft and spongy in texture.
- Velvet-like fungus mycelium and spores are produced on the lesion surface under cool, humid conditions.

***Courtesy*: Marita Cantwell**

Management

- Field sanitation - remove and burn decaying infected plant parts.
- Space seedlings and transplants to allow for free flow of air through the crop.
- Treatment with hot air at 38°C for h or hot water at 50°Cto 53°C for 2 to 3 min.

Late Blight

Pathogen: *Phytophthora capsici*

Symptoms

- Infected leaf tissue - wilted, light green or gray-green, later becoming tan to white and scalded in appearance with moisture, leaf spots have a water soaked border.
- Fruit rots - irregular in shape and olive green or light green with water soaked borders.
- Rots expand rapidly and fruits can be completely diseased and desiccated, causing the formation of "mummified" fruits.
- Infected seeds are brown and shriveled.

Courtesy: **Marita Cantwell**

Management

- Rotation with non-susceptible crops will reduce the amount of *Phytophthora capsici* surviving in soil.
- Fresh, clean seeds should be planted in new potting mix to establish healthy transplants.
- Monitor seedlings as well as the field and remove diseased plants as soon as they occur.

Chapter 20

Postharvest Diseases of Carrot and their Management

V.K. Parthiban, S. Vanitha and K. Prabakar

Department of Plant Pathology, Centre for Plant Protection Studies, Tamil Nadu Agricultural University, Coimbatore – 3, Tamil Nadu

Carrot (*Daucas carota* subsp. *sativus*) is a root vegetable and it belongs to the family Apiaceae. The roots contain high amount of alpha and beta-carotene, and are a good source of vitamin K and vitamin B6. Carrots are susceptible to a number of postharvest diseases and disorders, many of which can be controlled by good management in the field and in storage.This chapter clearly depicting about major postharvest diseases of carrot and their effective management practices.

Crown Rot

Pathogen: *Rhizoctonia solani*

Symptoms

- ✰ Damping-off of carrot seedlings and a crown rot later and during storage.
- ✰ Field symptoms include premature senescence and death of foliage.
- ✰ On carrot roots - dark brown sunken lesions or cankers near the crown or in other parts of the root – cavity spot.

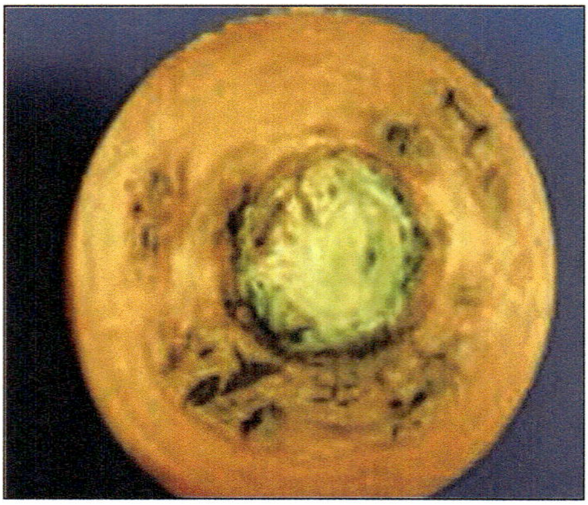

Initial symptom

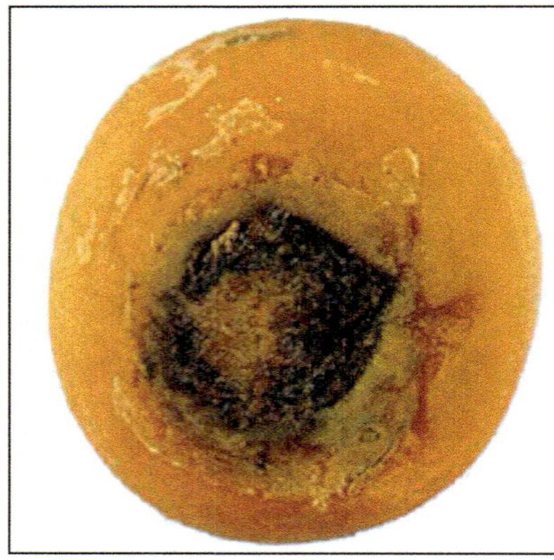

Dark brown sunken lesions

Courtesy: J.O. Strandberg

Management

- ✰ Cultural practices that reduce crown injury and enhance soil drainage and air circulation within the canopy are recommended.
- ✰ Plant debris from previous crops should be removed before planting carrots.
- ✰ Carrots should not follow perennial crops such as alfalfa.
- ✰ Quintozene – sprayed in the soil before sowing-effective control.

Cottony Soft Rot/White Mold

Pathogen: *Sclerotinia minor*

Symptoms

- ✰ In Foliage - water soaked, dark olive-green lesions associated with collapsed tissues.

Cottony white mycelium

Courtesy: Don Edwards

- Lesions expand rapidly over the entire leaf, petiole, and rosette with infected tissues covered by abundant cottony, white mycelium.
- Lesions of *Sclerotinia* rot advancing on carrot petioles and the rosette.
- At an advanced stage - affected tissues exhibit a bleached appearance, and occasionally an entire plant may collapse.

Management

- Storage - Good ventilation.
- Use of clean containers, maintenance of temp. near zero degree Celsius and RH - 95 per cent

Phytophthora Root Rot or Rubbery Brown Rot

Pathogen: *Phytophthora megasperma*

Symptoms

- Infected portions of the root become dark brown to black, water-soaked and rubbery in consistency.
- Lesions may occur in one or more bands anywhere on the carrot root.

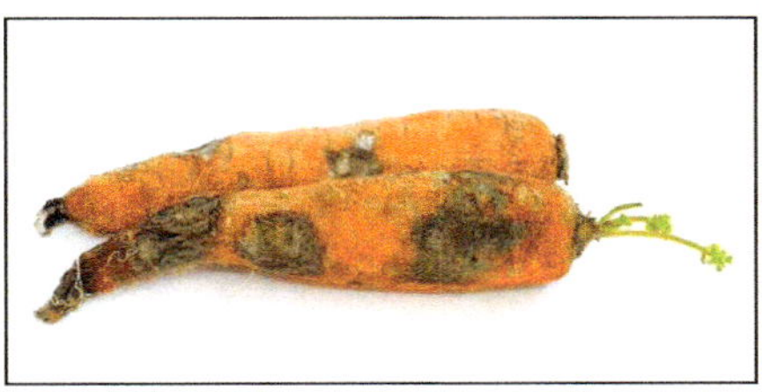

Dark black rings

Courtesy: **O. Lastochkina**

Management

- Maintaining adequate soil drainage.
- Avoid prolonged periods of water saturation.
- Storage conditions - temperature at 32° F and RH <95 per cent

Black Rot

Pathogen: *Alternaria radicina*

Symptoms

- ✰ Seedling infection results in pre emergence and post emergence damping-off.
- ✰ Older senescing petioles on mature plants are particularly susceptible to infection, provide an avenue for infection of the carrot crown, which appears as a black ring of decay where the petioles attach to the root (black crown).
- ✰ In Stored carrots - dry, black, sunken lesions which can decay the entire root and spread to adjacent carrots.

***Courtesy*: B. Raemart**

Management

- ✰ Application of Chlorothalonil by means of seed and soil application.
- ✰ Applying fungicides such as Iprodione and Strobilurins.
- ✰ Crop rotation with non host crops.
- ✰ Maintaining proper temperature and humidity (32°F and 95 per cent RH) in storage.

Bacterial Soft Rot

Pathogen: *Erwinia carotovora* subsp. *carotovora*

Symptoms

- ✰ Cells become water soaked, middle lamella is destroyed and the cells collapse.
- ✰ Soft, watery or slimy consistency.

- ☆ Rotted tissues – gray to brown, accompanied by foul odour.
- ☆ In the field, tops of rotted carrots turn yellow and wilt as roots break down.

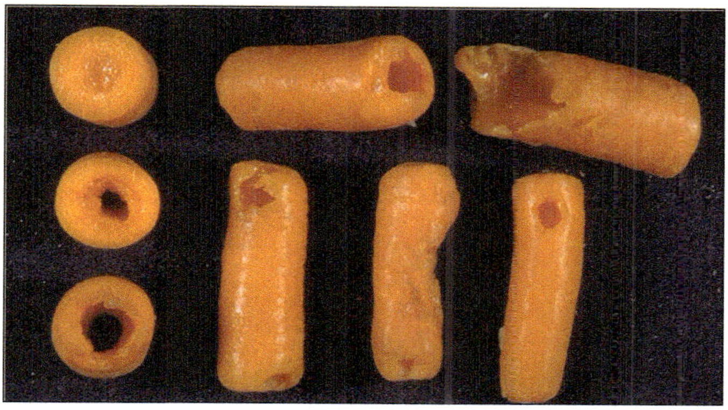

***Courtesy*: Don Edwards**

Management

- ☆ Careful handling at harvest - minimizes bruising of carrot roots.
- ☆ Washing - dipping in a solution of sodium hypochlorite (5.25 per cent).
- ☆ Dipping in a solution of 1:500 of sodium hypochlorite before storage or transits reduce the disease.
- ☆ Storage condition – temperature just above freezing point, RH 90 per cent.
- ☆ Crop rotation with alfalfa, beans, beets, corn.

Sour Rot

Pathogen: *Geotrichum candidum*

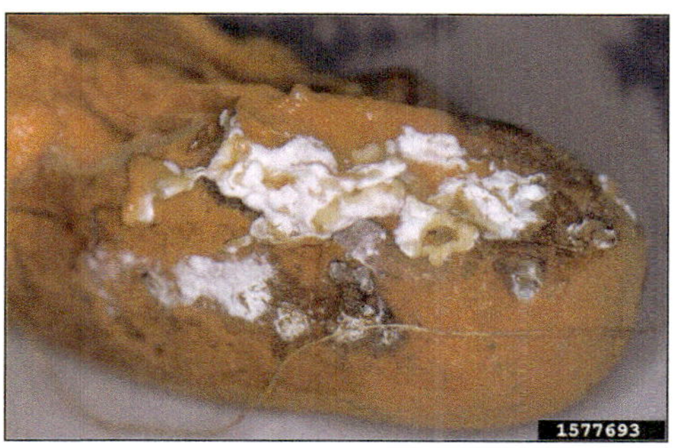

Soft watery colourless decay

***Courtesy*: J.O. Strandberg**

Symptoms

- Soft, watery, colorless decay on carrot roots.
- Decayed area - covered with dull, white spores of the pathogen and a vinegar-like odour may develop.

Management

- In the field, good field drainage minimizing wounding of carrots.
- In storage-good sanitation (use of new or disinfected storage containers), precooling and storing carrots at 32° F is essential.

Crater Rot

Pathogen: *Rhizoctonia carotae*

Symptoms

- Band of dark brown necrosis around the crown and horizontal brown canker-like lesions mostly on the crown and upper roots.
- Small pits subsequently developed beneath the lesions, that enlarged into sunken brown crater lines with a white, flocculent mycelium produced under high humidity.

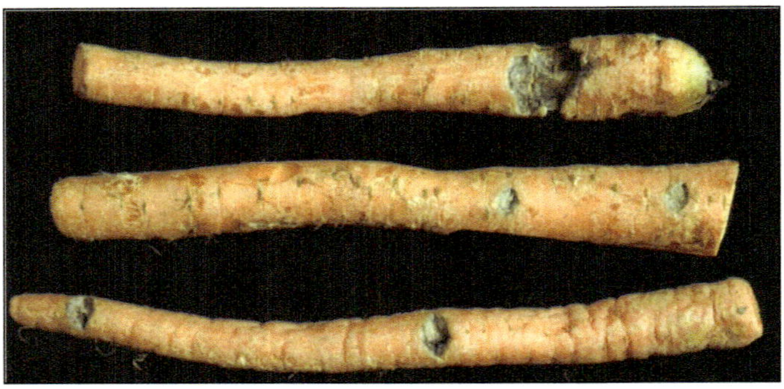

Courtesy: **Don Edwards**

Management

- Using disease free seed.
- Rapid removal of field heat.
- Eight years rotation with non host crop.
- Application of Dowicide (Ophenylphenol) @ 98 g/L.
- Application of Chlorothalonil reduces the disease incidence

Chapter 21

Postharvest Diseases of Potato and their Management

K. Prabakar, S. Vanitha and S. Parthasarathy

Department of Plant Pathology, Centre for Plant Protection Studies, Tamil Nadu Agricultural University, Coimbatore – 3, Tamil Nadu

Potato tuber (*Solanum tuberosum* L.) is the most and first consumed food crop in the world, because the dry matter, edible energy and edible protein content of potato makes it nutritionally higher vegetable as well as staple food throughout the world. Although, potato might be stored for longer time compared to other crops, tubers are often affected by some physiological and diseases disorders due often to poor handling and storage and losses are reported to vary between 40-50 per cent. These postharvest losses could be qualitative and quantitative losses. Beside sprouting which is the main important disorder affecting potato during storage, other physiological caused by environmental conditions could affect tubers and these are caused by exposure to:

1. Extreme temperatures before and during storage, overheating of tubers due to direct exposure to sunlight or non-refrigerated storage, and
2. Rough handling of tubers during harvesting. The pathological disorders might be cause by attacks of pathogens and insects and the diseases would be aggravated by poor or inappropriate handling and storage practices.

The aim of this chapter is to review the different postharvest diseases and disorders of potato tubers and how to reduce their impact by using classic and modern technologies to extend the shelf-life of tubers.

Tuber Rot

Pathogen: *Sclerotium rolfsii*

Symptoms

- Wilting is the initial symptom.
- Yellowish brown coloured sclerotia appeared on the infected tuber.
- Rotting of the tuber.
- Milky white and floccose appearance of the tuber.

Courtesy: **Brad Geary**

Management

- Cultural practices like heavy earthing and irrigation at regular intervals can also check the disease.
- The disease is low in the variety Kufri Sindhuri.
- Among the Indian commercial cultivars, Kufri Bahar, Kufri Chamatkar, Kufri Jyothi, Kufri muthu and Kufri swarna are resistant.
- The disease can be controlled to a certain extent by growing non susceptible crops like corn and sorghum.

Black Scurf

Pathogen: *Rhizoctonia solani*

Symptoms

- Black speck, black speck scab, russet scab on tubers.
- At the time of sprouting dark brown colour appear on the eyes.
- Affected xylem tissue causes to wilting of plants.
- Infected tuber contains russeting of the skin.
- Hard dry rot with browning on internal tissue.

- Spongy mass appear on the infected tuber. Seed tubers are source of spread.
- Moderately cool, wet weather and temp 23°C are the favourable for the development of disease.

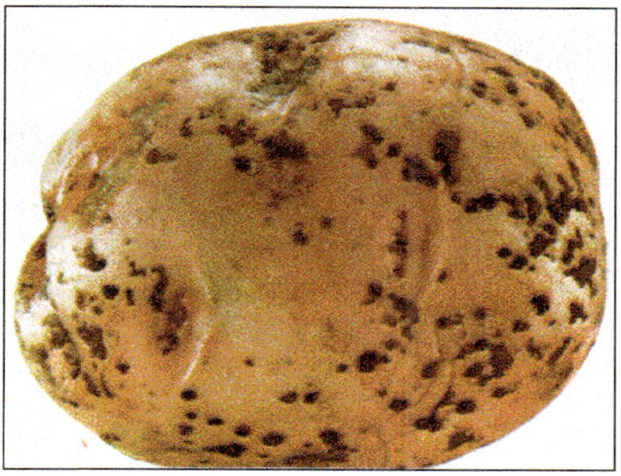

***Courtesy*: Dennis Johnson**

Management

- Disease free seed tubers alone should be planted.
- Treating the soil with Pentachloronitrobenzene at the rate of 70kg/ha lowers the incidence of the disease.
- Well sporulated tubers may be planted shallow to control disease.
- The disease severity is reduced in the land is left fallow for 2 years.

Common Scab or Corkey Scab

Pathogen: *Streptomyces scabies*

Symptoms

- Corkiness of the tuber periderm is the characteristics symptoms.
- ¼ inch into the tuber surface are russette appearance.
- Slightly pitted on the infected tuber.
- Light brown to dark brown lesion appears on the infected tuber.
- Affected tissue will attract insects.

Management

- Only scab free seed potatoes should be planted as this will help in checking the spread of the inoculum and infection to be subsequent crop.
- Infection of the seed tubers can be removed by 1.5 hrs dip in mercuric chloride 0.1 per cent solution or by 2h dip in 1 part formaldehyde in 240 parts of water.

***Courtesy*: Brad Geary**

- Four to six years crop roation with alfalfa satisfactory under irrigated conditions.
- The disease incidence can be effectively reduced by green manuring the fields before planting potatoes.
- Common scabis severe in alkaline soil and application of alkaline fertilizers like calcium ammonium nitrate should be avoided.

Brown Rot or Bangle Blight

Pathogen: *Ralstonia solanacearum*

Symptoms

- At the time tuber formation wilt is the main characteristic symptom.
- In leaf symptom - wilt, stunt and yellowing.
- Browning of xylem tissue.
- Eye buds are black in colour.
- Bacteria ooze coming on infected tuber surface and emits a foul odour.

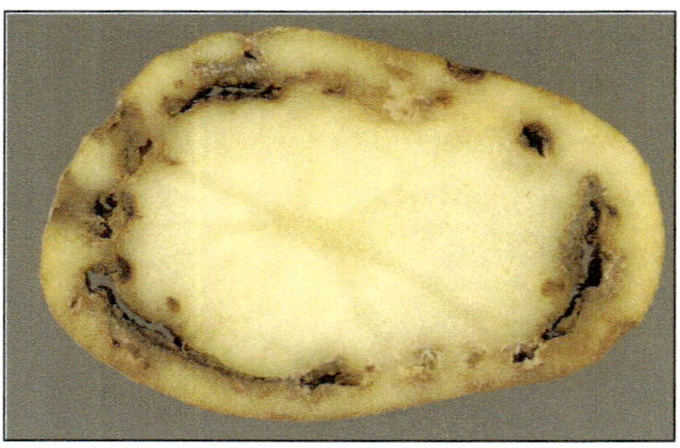

***Courtesy*: Brad Geary**

- Under favorable conditions, potato plants infected with *R. solanacearum* may not show any disease symptoms.
- In this case, latently infected tubers used for potato seed production may play a major role in spread of the bacterium from infected potato seed production sites to healthy potato -growing sites.
- *R. solanacearum* can survive for days to years in infected plant material in soils, infested surface irrigation water, infected weeds, and infected potato washings and sewage.
- From these sources of inoculum, bacteria can spread from infested to healthy fields by soil transfer on machinery, and surface runoff water after irrigation or rainfall.
- Infected semi-aquatic weeds may also play a major role in disseminating the pathogen by releasing bacteria from roots into irrigation water supplies.

***Courtesy*: Brad Geary**

Soft Rot

Pathogen: *Pectobacterium carotovorum* subsp. *carotovorum*

Symptoms

- Infection at two phases are black leg and soft rot.
- Black lesion appear on the base of the plant.
- Systemic and browning of infected tubers.
- Yellow appearance of the plant.
- Finally the plants wilt and die. Lenticels (water soaked brown rot).
- Rot and collapse of tubers.
- Soft, reddish or black ring appear on the infected tuber.

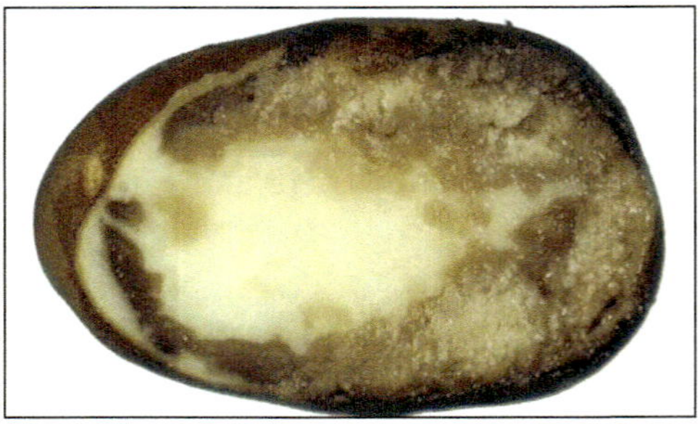

***Courtesy*: Brad Geary**

Management

- The pathogen is difficult to control because of long survival both on seed tubers and in soils.
- However using disease free seed tubers could minimize the disease incidence.
- Before planting the seed tubers are treated with Boric Acid (3 per cent for 30 minutes) and dried in shade.
- The same treatment is repeated before the storage of the tubers.
- Following crop rotations with wheat, pea, oats, barley, lupin, soybean, sorghum and bajra checks the disease development.
- In plains, treatment of the seed tubers with acetic acid + 0.05 per cent Zinc Sulphate solution or Carbendazim 1 per cent for 15 minutes effectively controls the disease.

Chapter 22

Postharvest Diseases of Beans and their Management

S. Parthasarathy and K. Prabakar

Department of Plant Pathology, Centre for Plant Protection Studies, Tamil Nadu Agricultural University, Coimbatore – 3, Tamil Nadu

Bean (*Phaseolus vulgaris* L.) is one of the most important economic vegetable crop in Bean (*Phaseolus vulgaris* L.) is one of the most important economic vegetable crop in Egypt for both local consumption and exportation for Europe and other countries. Snap bean pods could be attacked with several diseases caused by fungi during the growth in the fields, harvest, storage, marketing or exporting. Many phytopathogenic postharvest fungi such as *Sclerotinia sclerotiorum, Botrytis cinerea, Alternaria alternata, Rihzctonia solani, Pythium aphanidermatum* and *Fusarium solani* cause great losses in quantity and quality of the snap bean pods. These fungal pathogens are field-borne and may cause disease in the field to reduce yield. However, due to latent infection they frequently are troublesome in storage. Therefore, it is essential to correctly manage disease in the field prior to harvest. The most serious postharvest diseases were gray mould (*Botrytis cinerea*) and white mould (*Sclerotinia sclerotiorum*).

Anthracnose

Pathogen: *Colletotrichum lindemuthianum*

Symptoms

☆ Bean pods with black, sunken lesions or reddish –brown blotches most likely have anthracnose, a fungal disease caused by *Colletotrichum lindemuthianum*.

- Black, sunken lesions about ½ inch in diameter develop on stems, pods and seedling leaves (cotyledons) but are most prominent on pods.
- A salmon colored ooze on lesions and the veins on lower leaf surfaces turns black.
- On lima beans, symptoms are sooty – appearing spots on leaves and pods.
- Anthracnose develops primarily during the spring and fall when the weather is cool and wet, and not during our hot, dry summers.
- Lima beans are particularly susceptible.

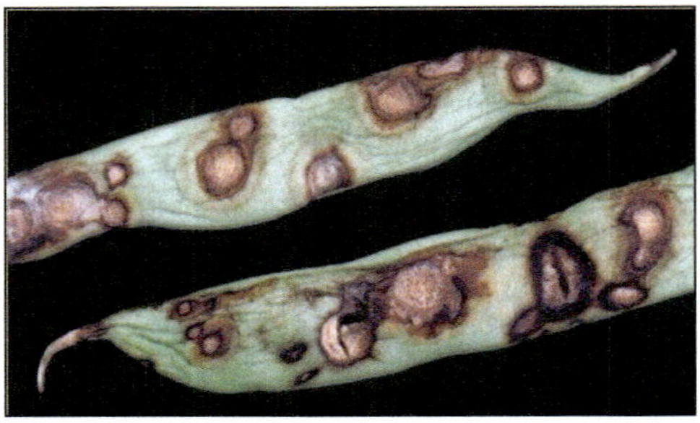

***Courtesy*: Shrinkala Manandhar**

Management

- Prevent this disease by using certified disease – free seed for planting and removing all plant debris after harvest.
- Anthracnose can survive in the soil for two years on plant debris or be brought to the garden on infected seeds.
- Do not plant bean seeds in an area that had disease for two to three years.
- Avoid over head watering and avoid splashing soil on to the plants when watering.
- Fungicide sprays of copper are the only recommended chemical that can be used on lima beans for anthracnose control.

Chapter

23

Postharvest Diseases of Onion and their Management

M. Karthikeyan and K. Prabakar

Department of Plant Pathology, Centre for Plant Protection Studies, Tamil Nadu Agricultural University, Coimbatore – 3, Tamil Nadu

Onions (*Allium cepa* L.) are an important seasonal crop in the world, and they are one of the commercially valuable crop. Therefore, the long-term storage of onions is necessary to ensure a steady supply throughout the year. The main cause for the reduction in onion yields is the various fungal infections that appear during long-term storage. It is vital that onion diseases are diagnosed early and effective management strategies are implemented to reduce the loss of onions. Onion quality is associated with many factors such as disease and humidity, *etc.* during cultivation, production, storage, and distribution. Despite the advances in production technology, the postharvest loss during storage is still a major problem. Onions suffer from many postharvest diseases such as black mold, blue mold, neck rot, brown rot, soft rot, and smudge, among which, black mold, blue mold, and gray mold are the predominant diseases that restrict the domestic and international trade of onions.

Smut

Pathogen: *Urocystis cepulae*

Symptoms

- ✰ Black smut sori are seen at the base of the leaves and leaf surface.

- Black powdery mass is seen after rupturing of sorus wall.

Courtesy: **Howard F. Schwartz**

Management

- Seed treatment with Thiram or Captan @4g/kg.
- The bulbs may be dipped in Thiram solution 0.25 per cent.

White Rot

Pathogen: *Sclerotium cepivorum*

Symptoms

- The leaves become yellow and die -back and when the plants are pulled up, roots are found to be rotten and the base of the bulb covered with a white or gray fungal growth.
- Later, numerous small black spherical sclerotia are produced.
- The bulb of the onion completely rots.

Courtesy: **Ralph Noble**

Management

- Crop rotation and clean seed are the only effective control.
- Heavy manuring with organic manures reduces the disease in the crop.
- Seed dressing with Benomyl, Carbendazim or Thiophanate - methyl (100 to 150 g/kg seed) gives effective control.

Purple Blotch

Pathogen: *Alternaria porri*

Symptoms

- This disease occurs mainly at the top of the leaves, the infection starts with whitish minute dots on the leaves with irregular chlorotic areas on tip portion of the leaves.
- Circular to oblong concentric black velvety rings appear in the chlorotic area.
- The lesions develop towards the base of the leaf.
- The spots join together and spread quickly to the entire leaf area.
- The leaves gradually die from the tip downwards.

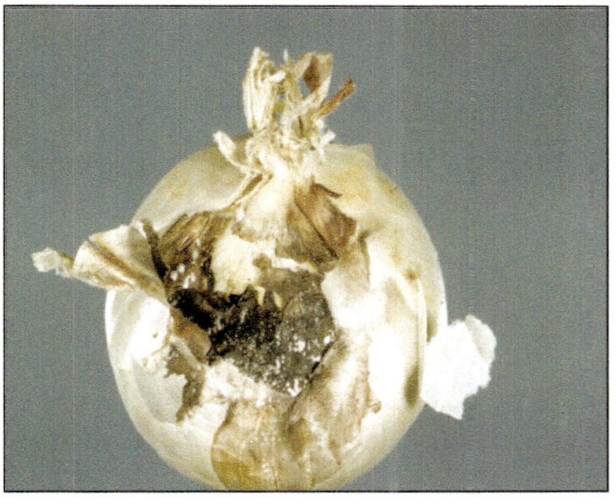

***Courtesy*: Howard F. Schwartz**

Management

- Disease free bulb should be selected for planting.
- Seeds should be treated with Thiram @4g/kg seed.
- The field should be well drained.
- Three foliar sprayings with Copper oxy chloride 0.25 per cent or Chlorothalonil 0.2 per cent or Zineb 0.2 per cent or Mancozeb0.2 per cent.

Neck Rot

Pathogen: *Botrytis allii, B. squamosa and B. cinerea*

Symptoms

- ☆ Symptoms usually appear after harvest, although infections originate in the field.
- ☆ Greatest epidemic development occurs when cool (50°to75°F), moist weather prevails for some days before or during harvest.
- ☆ If the weather remains dry during harvest and curing, losses found in storage are usually small.
- ☆ Symptoms are first seen as a softening of the tissues around the neck of the bulb, or more rarely, at a wound.
- ☆ A definite margin separates diseased and healthy tissues.
- ☆ Infected tissues become sunken, soft, and appear brownish to grayish in color, as if they had been cooked.
- ☆ These symptoms progress gradually to the base of the bulb.
- ☆ Then the entire bulb may be come mummified.
- ☆ Hard, irregularly shaped kernel-like bodies, sclerotia, may form between scales, especially at the neck region.

***Courtesy*: Dr. Steven Roberts**

Blue Mould Rot

Pathogen: *Penicillium* sp.

Symptoms

- ☆ Blue mold generally appears during harvesting and storage.
- ☆ Initial symptoms include water soaked areas on the outer surface of scales.

- Later, a green to blue green, powdery mold may develop on the surface of the lesions.
- Infected areas of fleshy scales are tan or gray when cut.
- In advanced stages, infected bulbs may disintegrate into a watery rot.
- Many species of *Penicillium* can cause blue mold.
- These fungi are common saprophytes on plant debris and senescent plant tissue.

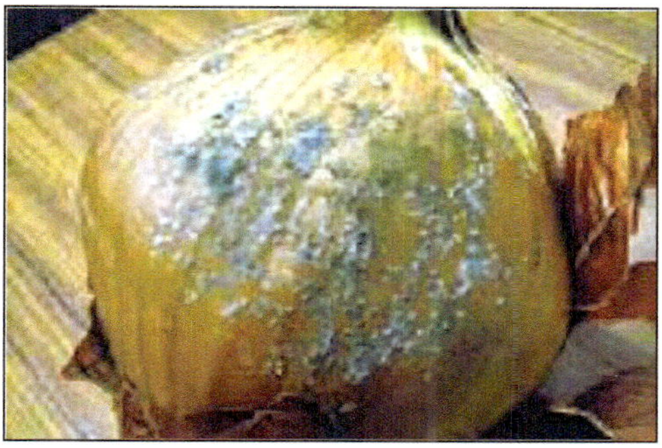

***Courtesy*: TNAU Portal**

Black Mould

Pathogen: *Aspergillus niger*

Symptoms

- Infection usually is through neck tissues as foliage dies down at maturity.
- Infected bulbs are discolored black around the neck, and affected scales shrivel.

***Courtesy*: Howard F. Schwartz**

- Masses of powdery black spores generally are arranged as streaks along veins on and between outer dry scales.
- Infection may advance from the neck into the central fleshy scales.
- In advanced disease stages, the entire bulb surface turns black, and secondary bacterial soft rot may make the bulb soft and mushy.
- No external symptoms may be found with some bulbs.

Management

- Seeds should be treated with Thiram @ 4g/kg seed.
- The field should be well drained.
- Three foliar sprayings with Copper oxy chloride 0.25 per cent or Chlorothalonil 0.2 per cent or Zineb 0.2 per cent or Mancozeb 0.2 per cent.
- Growers must follow crop rotation and harvested bulbs must be thoroughly cured to reduce potential storage losses.
- Soil drenching with Copper oxy chloride 0.25 per cent.

Chapter 24

Postharvest Diseases of Garlic and their Management

K. Prabakar, S. Vanitha and S. Parthasarathy

Department of Plant Pathology, Centre for Plant Protection Studies, Tamil Nadu Agricultural University, Coimbatore – 3, Tamil Nadu

Allium sativum (Liliaceae), known as garlic, is a strongly aromatic bulb crop. Garlic is one of the most popular spices in the world. Garlic grows in temperate and tropical regions all over the world, and many cultivars have been developed to suit different climates. In India, garlic is used to relieve problems, such as coughs and fevers or applied externally to prevent graying of hair and to improve skin conditions, such as eczema and scabies as well as to treat tetanus and lungs inflammation. But they are attacked by various pathogens after harvest and deteriorate its market value. They are field-borne and may cause disease in the field to reduce yield. However, due to latent infection they frequently are troublesome in storage. Therefore, it is essential to correctly manage disease in the field prior to harvest.

White Rot

Pathogen: *Sclerotium cepivorum, S. rolfsii*

Symptoms

- Leaves decay at the base, turn yellow, wilt, and topple over.
- Roots and bulbs - covered with a fluffy white mycelium.

- Affected bulbs may become watery, and the outer scales crack as the bulb dries and shrinks.
- Micro sclerotia of the White Rot fungus *S.cepivorum* developing on an infected garlic bulb.

***Courtesy*: Ralph Noble**

Management

- Rotating out of Allium crops for ten years.
- Destroying infected tissue.
- Planting disease - free seed stock.
- Seed dressing with benomyl or carbendazim (4g/kg of seed)
- Micro sclerotia of the White Rot fungus *S.cepivorum* developing on an infected garlic bulb.

Basal Rot

Pathogen: *Fusarium oxysporum* f. sp. *garlic*

Symptoms

- Plants - show reduced emergence, yellowing and/or browning (necrosis) of leaves beginning at tips.
- Reduced bulb size, bulb decay, and brown, poorly developed root systems.
- In storage - bulbs show spongy, sunken, yellow brown rotting lesions.

- ✰ In early stages - infected bulbs are softened, brown and watery when cut open white, light pink or reddish fungal growth covering the cloves or in rot cavities.
- ✰ Deep cracks form in the cloves, followed by break down of the tissue, which will eventually dry down to a portion of its original size, the cloves becoming crinkled and small.

***Courtesy*: Howard F. Schwartz**

Management

- ✰ Avoid rotations with *Allium* spp. (*e.g.* onions and leeks) and cereals.
- ✰ Store bulbs at cool temperatures and low humidity with good ventilation.
- ✰ Avoid storing damaged bulbs.

Blue Mould Rot

Pathogen: *Penicillium* spp.

Symptoms

- ✰ A blue-green color powdery mould is observed on cloves in soil and in storage, thus its common name, "Blue Mold".

***Courtesy*: S.K. Mohan**

Management

- Bulbs are harvested carefully to avoid wounds and bruising, then promptly dried or cured.

Black Mould

Pathogen: *Aspergillus niger* and *A. alliaceous*

Symptoms

- Whole tissues become black powdery mass.
- Individual bulbs shrivel and become light in weight.
- Under high humid condition the inner tissues become moderately soft.
- Infected bulbs lose their pungency and smell.
- Rotten garlic cloves show black, brown, pink or white coloured rotting.

***Courtesy*: Howard F. Schwartz**

Management

- Rapid and thorough curing.
- Storage condition should be with good ventilation and temperature just above 0°C.

Pink Rot

Pathogen: *Pyrenochaeta terrestis*

Symptoms

- Roots are affected and they turn pink or reddish and sometimes dark to a red or purple colour.
- Black spores form on the diseased roots which eventually shrivel and die.

Red colour roots

Courtesy: **Howard F. Schwartz**

Internal Bulb Rot

Pathogen: *Macrophomina phaseolina*

Symptoms

- No external symptom can be observed unless the outer scales are removed.
- Black pin head microsclerotia develop over the fleshy scales which are light in weight, brown, mostly shriveled.
- Do not lose their pungency.

Courtesy: **Howard F. Schwartz**

Management

- Dipping of bulbs in formalin 0.03 per cent Boric acid-2.0 per cent. After harvest to minimize storage rot.

Neck Rot

Pathogen: *Botrytis allii*

Symptoms

- Found upon the bulbs at the time of harvest.
- Affected scale tissue become soft and dense layer of gray mould appear at the neck.
- Infection progresses most rapidly down the scales which have been originally infected.

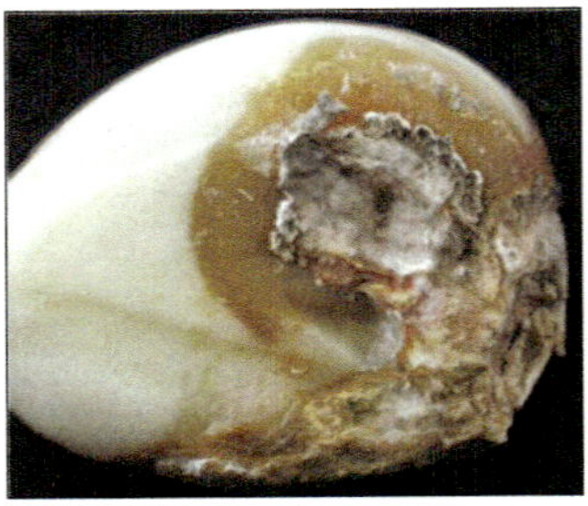

Rotten neck

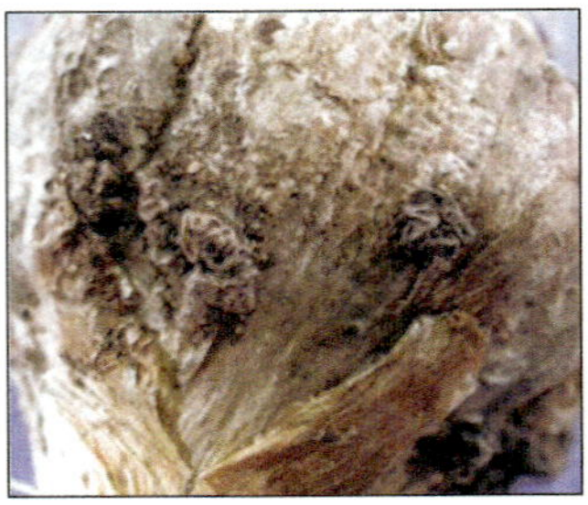

Gray mould on clove

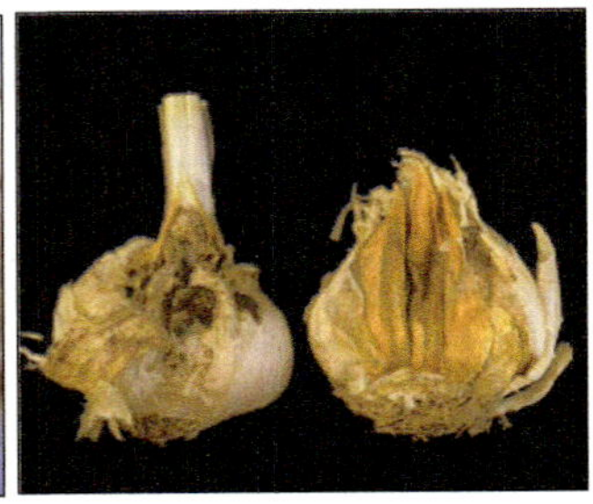

Black sclerotia on bulb

Courtesy: **Steven Roberts**

Management

- Promoting rapid drying at harvest and good aeration in storage is best for managing Botrytis on bulbs.
- Additionally, cooler storage temperatures may help control the disease

Chapter

25

Postharvest Diseases of Cucurbits and their Management

K. Kavitha, S. Vanitha and K. Prabakar

Department of Plant Pathology, Centre for Plant Protection Studies, Tamil Nadu Agricultural University, Coimbatore – 3, Tamil Nadu

Cucurbits are the popular name of the family cucurbitaceae, commonly known as the gourd family. They are widely distributed in the tropics and warm temperate regions of south, southeast and East Asia, Africa including Madagascar, central and south America. The family is represented by about 120 genera and 800 species. Cucurbits are mostly climbers and trailers, are rarely woody and arborescent. They are characterized by inferior ovary and parietal placentation. The most common uses of cucurbits are as vegetables and fruits. They are valuable sources of vitamins and minerals. River bed cultivation of a number of cucurbits is economical. The river beds of big rivers used for cucurbit cultivars are: Ganga, Yamuna, Saraswati, Narmada, Sutlej, Krishna, Kaveri, Godavari, Mahanadhi, Sabarmati, Bramhaputra *etc.* Cucurbits are popular because they are easy to grow. They do not require high inputs. They can be grown on even soils of marginal fertility. They are useful to prevent soil erosion.The cucurbits are not usually processed. Similarly muskmelon is not generally processed except the flesh of muskmelon is used in ice-cream in western countries. On small scale the juice of muskmelon flesh could be prepared; however, it has to be consumed within short time because it does not have any storage quality.

Anthracnose

Pathogen: *Colletotrichum orbiculare*

Other crops like Watermelon, Cantaloupe, Cucumber, Squash and pumpkin are also affected by this pathogen.

Symptoms

- ☆ Lesions on watermelon are irregular and turn dark brown or black.
- ☆ Immature fruit - turn black, shrivel, and die.
- ☆ Round, water-soaked spots develop on the older fruit.
- ☆ Spots turn dark green to brown with age and may become sunken.
- ☆ Under wet conditions, pinkish-colored spore masses can be seen oozing out of the sunken spots.
- ☆ Lesion on watermelon showing a gelatinous mass of salmon colored spores.
- ☆ Melon showing the blackened center of the lesion and a hint of the pinkish spore mass.

Courtesy: **Dominique Blancard**

Management

- ☆ Field sanitation - destroy the plant debris.
- ☆ Hot water treatment of 57.2°C for 20 min.
- ☆ Seed treatment - Thiram or Carbendazim or 2g/kg.
- ☆ Spraying at weekly intervals of Carbendazim 0.1 per cent, Mancozeb 0.2 per cent, Difolaton 0.2 per cent. Fruit dip - 5 min in wash water containing 120 ppm of chlorine helps to prevent infection of healthy fruits.
- ☆ Resistant varieties in watermelon - Black Stone, Congo, Diamond, Charleston.

Alternaria Fruit Rot

Pathogen: *Alternaria alternate* f. sp *cucurbitae*

Crop: Melon

Symptoms

- ☆ Small dark shallow circular lesion that can occur anywhere on the fruit surface.
- ☆ The lesion can be extend into the fruit flesh in some cases light gray to black mycelium will grow abundantly over the melon surface.

Courtesy: **Anthony Keinath**

Management

- ☆ Proper sanitary practices.
- ☆ Crop rotation.
- ☆ Spraying of fungicides *viz.*, Azoxystrobin 0.1 per cent, Chlorothalanil 0.1 per cent and Copper oxy chloride 2.5 g/liter of water.

Pumpkin Black Rot

Pathogen: *Didymella bryoniae*

Symptoms

- ☆ The surface of a stored pumpkin, showing developing rot.
- ☆ Two sunken lesions are visible (centre, to upper right), that have developed radially from the point of infection, dissolving the skin of the pumpkin to produce a granular surface showing rings.
- ☆ The infection also softens and liquefies the internal tissues of the pumpkin.
- ☆ *D. bryoniae* is the most common cause of rotting of stored pumpkins, and presents a variety of appearances, including the eventual development of large lesions consisting of softened black skin on the fruit.

- ✰ Storage rots in pumpkins may be caused by premature harvesting, and mechanical or insect damage to the skin.
- ✰ The risk of rot is also greater in fruits stored at temperatures below 10 degrees centigrade, and by exposure to high humidity.

Courtesy: **Gerald Holmes**

- ✰ Affected area - brownish and water soaked. Advanced stages - rind becomes black and deeply wrinkled.
- ✰ Large irregular areas of the fruit become bronzed with distinct concentric rings.

Management

- ✰ Use of disease-free seed.
- ✰ Two years crop rotation out of all cucurbits.
- ✰ Field sanitation.
- ✰ Application of fungicides *viz.*, Chlorothalanil, Mancozeb and Benomyl.
- ✰ Cucumbers - precooled to 10°C or lower temperature.

Coanephora Wet Rot

Pathogen: *Coanephora cucurbitarum*

Crop: Squash

Symptoms

- ✰ Attacks the blossoms first and progresses into the developing fruit causing a wet rot at the blossom end.

- Fruit rot progresses rapidly and can affect entire fruit within one or two days.
- Sporulation by the fungus appears as spines with dark heads on the surface of infected tissues.

Courtesy: **John Hartman**

Management

- Crop management practices, reduce soil moisture (raised beds) and prevent fruit injury.
- Prevent soil contact with the soil (plastic mulches or trellising).
- Postharvest losses may be reduced by harvesting fruits at proper stage of maturity.
- Minimizing cucurbit fruit injuries at harvest.
- Pre cooling fruit.
- Maintaining relatively low storage temperature.

Cottony Leak

Pathogen: *Pythium aphanidermatum, P. debaryanum, P. ultimum*

Crop: Cucumber, Squash, Watermelon

Symptoms

- Pythium fruit rot symptoms begin as brownish, water-soaked lesions that quickly become large, watery, soft, and rotten.
- The rot generally begins on the parts of fruit in contact with the soil.
- In cucumber, a brown to dark green blister often is apparent on fruit before they become watery and rot.
- White cottony mycelium is apparent on rotten tissues, especially during humid weather.

- *Pythium* fruit rot is most severe in poorly-drained fields during wet weather.
- The disease can render fruit unmarketable.

Courtesy: **Cheryl Kaiser**

Management

- Soil drenching with copperoxychloride 0.25 per cent.
- Fruits should be kept away from soil.

Belly Rot

Pathogen: *Rhizoctonia solani*

Symptoms

- Dark brown water-soaked decay on the side of the fruit in contact with the soil.
- Followed by a yellowish-brown discolouration of the fruit surface. Entire fruit rot within few days.
- Water-soaked lesions.

Courtesy: **Cheryl Kaiser**

Management

- ☆ Pre-harvest sprays of the fungicides *viz.,* Azoxystrobin, Chlorothalonil, Thiophanate-methyl.
- ☆ Holding the fruit at 10°C (50°F) will retard disease development during transit and storage.

Lasiodiplodia Fruit Rot

Symptoms: *Lasiodiplodia theobromae*

Crop: Cantaloupe, Watermelon

Symptoms

- ☆ Water soaking is seen at the stem end and the tissue is soft.
- ☆ The stem end of the fruit shrivels and turns brown.
- ☆ The fruit first appears water soaked and later darkened and shrivels.
- ☆ When harvesting, a large portion of the peduncle should remain attached to the fruit.

***Courtesy*: Julie Beale**

Management

- ☆ Scratches and bruises must be avoided.
- ☆ Pre cooling after harvest.
- ☆ Harvest fruits with long stems.
- ☆ Cut ends painted with a fungicide paste - copper sulphate.

Bacterial Soft Rot

Pathogen: *Erwinia carotovora*

Symptoms

- ☆ Infects the fruit via cracks or wounds in the skin.

- Soft rot rapidly disintegrates the flesh, turning it into a soft mass of leaky tissue.
- Infected fruits typically have a foul odour.

Courtesy: **Tom Isakeit**

Management

- Avoid injury to the skin. Use properly sanitized (*i.e.* 150 ppm hypochlorous acid) wash water.

Bacterial Fruit Rot

Pathogen: *Xanthomonas campestris* pv. *cucurbitae*

Symptoms

- Fruit - small, slightly sunken, circular spots with a tan center and dark brown border.
- Epidermis may split, spots enlarge, and become sunken.
- Bacteria can penetrate into the flesh causing fruit rot and other secondary bacteria may invade.

Courtesy: **Payton Strawser**

Management

- ☆ Seed treatments with hot water (50°C for twenty minutes) or 10 per cent Chlorox.
- ☆ Avoid overhead irrigation and working the fields when they are wet.
- ☆ Rotate out of cucurbits for two years.
- ☆ Repeated applications of copper fungicides as a protectant may be helpful.

Phytophthora Rot

Pathogen: *Phytophthora capsici*

Crop: Cucumber, Watermelon, Cantaloupe

Symptoms

- ☆ *Phytophthora* blight symptoms may be observed on all above-ground parts of a cucurbit plant.
- ☆ However, cucumbers and watermelons usually have symptoms on fruit, but not on foliage.
- ☆ Soft sunken spots on fruit surface.
- ☆ Under conditions of high humidity white mycelium may colonize the fruit surface. Lesions on fruit vary depending on the host.
- ☆ Lesions on watermelon are often round, water-soaked and may appear as a bruise.
- ☆ Under moist conditions, these lesions are covered with a white mold, caused the growth of *P. capsici*.
- ☆ Lesions on fruit may be more common on the underside of the fruit where moisture accumulates.
- ☆ Fruit lesions on pumpkin may be large and occur in no particular shape.

***Courtesy*: Payton Strawser**

Management

- Rotation with non-host crops is recommended.
- Avoid growing alternate hosts such as pepper, tomato, eggplant, cocoa, and macadamia.
- Manage soil moisture by selecting well-drained fields, avoiding low-lying areas, sub soiling, preparing dome-shaped raised beds for non-vining crops, and not over irrigating.
- Avoid to take soil from infected fields.

Chapter 26

Postharvest Diseases of Crops under Protected Cultivation and their Management

N. Indra

Regional Research Station, Tamil Nadu Agricultural University, Paiyur, Krishnagiri Dt. – 635 112, Tamil Nadu

I. STRAWBERRY

Strawberry fruits are particularly perishable, especially after harvest. Even if they are apparently healthy at the time of harvest, they can undergo spoilage. The main causes of decay of strawberry fruit during storage and shelf life are the development of rots that are caused by a range of fungi. This chapter clearly explains about major postharvest diseases of crops under protected cultivation and their management strategies.

Rhizopus Soft Rot

Pathogen: *Rhizopus stolonifer*

Symptoms

- ☆ Initial infections appear as discoloured, water-soaked spots on fruit.
- ☆ Due to the enzymatic breakdown, the lesions enlarge rapidly and cause the berry limp brown, and the contents leak out onto the bed.

- Under conditions of high relative humidity, the berry rapidly becomes covered with a coat of white mycelium and sporangiophores.

Rhizopus soft rot

Courtesy: **Steven Koike**

Management

- Discarded plant refuses or berries in the furrows.
- Remove all ripe fruit from the field. Avoid wounding of fruits during harvest.
- While picking ensure that the entire fruit is removed from the stem, not leaving behind the fleshy receptacle of the fruit as it can serve as a site for invasion by fungus.
- *Rhizopus* stops growing at temperatures below 8° to 10°C, so rapid postharvest cooling of fruit is essential for disease control.

Botrytis Fruit Rot (Gray Mold)

Pathogen: *Botrytis cinerea*

Symptoms

- Young blossoms are very susceptible to infection.
- One to several blossoms in a cluster may show blasting (browning and drying) that may spread down the pedicel. Fruit infections usually appear as soft, light brown, rapidly enlarging areas.
- If it remains on the plant, the berry usually dries up, "mummifies", and becomes covered with a gray, dusty powder.
- Fruit infection is most severe in well-protected, shaded areas of the plant where the humidity is higher and air movement is reduced.
- Berries resting on soil or touching another decayed berry or a dead leaf in dense foliage are most commonly affected.

- The disease may develop on young (green) fruits, but symptoms are more common as they mature.
- After picking, mature fruits are extremely susceptible to gray mold, especially if bruised or wounded.
- Under favorable conditions for disease development, healthy berries may become a rotted mass within 48 hours after picking.

  Note: The symptoms usually develop first on the calyx end of the fruit.

Botrytis fruit rot (gray mold)

***Courtesy*: Steven Koike**

Management

- Provide air circulation between plants and keep the leaves, flowers, and fruit dry.
- Select a well-drained site.
- Provide wide space between the rows, and avoid planting in excessively wide rows.
- Control weeds, as extra plants in the area will inhibit drying.
- Avoid excessive nitrogen application.
- Avoid overhead irrigation, which leaves foliage and fruit damp.
- Remove and destroy dead or infected plant material.
- Remove all ripe fruit during harvest and the decayed and rain damaged fruits.
- Remove all fruits after spring and summer rains.
- Use varieties that are suited to the growing area with necked fruit or reflexed calyx.
- Grow strawberries in plastic tunnels to effectively reduce the incidence of *Botrytis* infections.

- ☆ Use plastic mulches to prevent berry-soil contact.
- ☆ Refrigerate berries immediately to 35 to 40°F in order to slow the development of gray mold (*Botrytis*) and other fruit rots.

Anthracnose

Pathogen: *Colletotrichum acutatum*

Symptoms

- ☆ Affected plants show lesions that girdle the stems, causing individual leaves or entire plants to wilt.
- ☆ Under warm, humid conditions, salmon-colored masses of spores form on the lesions. When crown tissue is infected, the entire plant may wilt and die.
- ☆ When infected crowns are sliced open, internal tissue is firm and reddish brown.
- ☆ Crown tissue may be uniformly discolored or streaked with brown.
- ☆ On fruit, symptoms first appear as whitish, water soaked lesions up to 1/8 inch in diameter which turn brown and enlarge within 2 to 3 days to involve most of the fruit.
- ☆ Lesions are covered with salmon-colored spore masses.
- ☆ Infected fruit eventually dry down to form hard, black, shriveled mummies.
- ☆ Fruit can be infected at any stage of development.

Anthrocnose lesion on strawberry

***Courtesy*: Garrett Ridge**

Management

- ☆ Remove and destroy the infected plants.

- Avoid excess overhead irrigation.
- Do not over water or over fertilize.
- Use calcium nitrate source of nitrogen instead of ammonium

Leather Rot

Pathogen: *Phytophthora cactorum*

Symptoms

- On green fruit, the diseased areas remain firm.
- The green color may remain with brown margins around infected areas, but eventually the entire fruit becomes brown with a rough and leathery texture.
- Infected mature fruit have a bitter, unpleasant taste and odor even though there may be little change in color.
- Sometimes the berry has a brown to dark purple hue, and when sectioned, the vascular tissue to each seed is darkened.
- Under conditions of high humidity, white mycelia can grow on the surface of the fruit.
- Eventually, infected green and mature fruit become dry, shrivel, and mummify.
- After harvest white fuzzy growth may appear under moist packaging conditions.

Leather rot

***Courtesy*: Frank J. Louws**

Management

- Select a planting site that is well drained with good sun exposure.

- Use raised beds and proper row orientation to facilitate drainage.
- Do not allow soil to compact or develop ruts as it will encourage water puddling.
- Practice proper plant spacing and weed control to promote airflow for quick drying of plant surfaces.
- Use mulches such as straw or plastic mulch, to help reduce contact between fruit and soil, to provide a barrier between the plant and standing water, and to reduce rain splash.

II. CAPSICUM

- Capsicum is an important cash crop with susceptibility to a range of postharvest diseases including anthracnose caused by *Colletotrichum capsici*, soft rot caused by *Erwinia carotovora,* and infection by *Botrytis cinerea* and *Alternaria* spp.
- The infection not only causes severe economic losses for producers but also leads to a risk of food safety for consumers

Bacterial Soft Rot

Pathogen: *Erwinia carotovora* subsp. *carotovora*

Symptoms

- Soft rot begins in the peduncle and calyx tissue of harvested fruit.
- Internal tissues near the site of infection softens and the lesions rapidly expands reducing the interior of the pod to a watery mass within few days.
- Initially, the lesions on the fruit are light to dark-colored, water-soaked, and somewhat sunken.
- Fruits infected on the plant often collapse and hang on the plant like a water-filled bag.

Soft tissue on peduncle

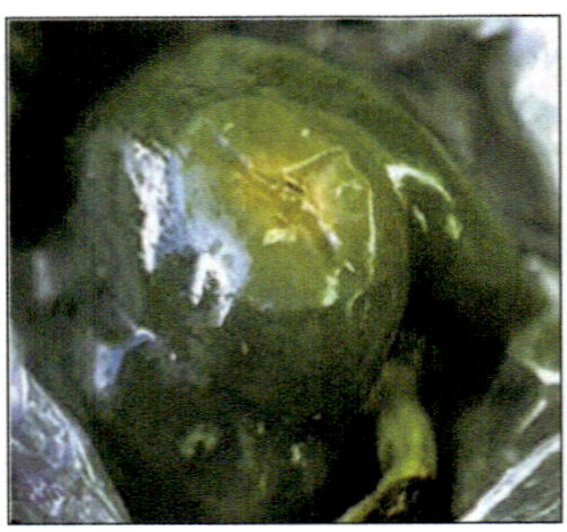

Water soaked sunken spots

Rotten chilli

***Courtesy*: Cynthia M. Ocamb**

Management

- ☆ Storage warehouses should be removed of all plant debris, and the walls and floors disinfected with either formaldehyde or copper sulphate between harvests.
- ☆ Injury to plant tissues should be avoided as much as possible, and the humidity and temperature of the storage facility should be kept low using an adequate ventilation system. Plants should be grown in well-drained soils.
- ☆ The control of specific insect vectors is also a good way of controlling disease spread in the field and in storage.
- ☆ Soil and foliage insecticide treatment helps controls the bugs that frequently cause wounds and disseminate the bacteria.
- ☆ Use chlorinated water to reduce populations of soft rot bacteria and to reduce the risk of infection during washing.
- ☆ Allow fruit to dry thoroughly.
- ☆ During packing and storage, the fruit should be kept clean and maintained in a cool, dry place.

Gray Mould

Pathogen: *Botrytis cinerea*

Symptoms

- ☆ The common symptom is the sudden collapse of succulent tissues such as young leaves, stems and flowers.
- ☆ The lesions rapidly expand forming irregular water soaked areas that result in death of seedlings, branches and older plants.

Powdery gray fungus on the fruit shoulders

***Courtesy*: Cynthia M. Ocamb**

- ✰ Gray powdery mass of fungal spores occur on the surface of the dead plant tissues.
- ✰ Fruit infections commonly begin at the calyx or at wound sites as soft olive green spots that covers the entire plant.

Management

- ✰ Reduce humidity in green houses and shade houses.
- ✰ Destroy crop residues after harvest and remove diseased fruits from packing sheds.
- ✰ Avoid wounding or damaging the fruit during harvest and handling.

Anthracnose

Pathogen: *Colletotrichum acutatum, C. gloeosporioides, C. capsici, C. coccodes*

Symptoms

- ✰ Characteristic symptoms appear on the mature pods as small water soaked sunken lesions that rapidly expand.
- ✰ The lesion increases in size and becomes sunken.
- ✰ The lesions are dark red to light tan in colour with stromatic fungal tissues.
- ✰ Pale to salmon spore masses occur scattered or in concentric rings on the lesion.

Sunken spots on mature fruits

Courtesy: **T.A. Zitter**

Management

- ✰ Use pathogen free seeds.
- ✰ Hot water treatment is recommended to destroy seed borne fungi.
- ✰ Soak seeds at 52°C for 30 minutes then plunge the hot seeds in cold water, dry on paper and dust with Thiram.

- ☆ Mulch to reduce soil splash on fruit and lower leaves.
- ☆ Avoid overhead irrigation.
- ☆ Harvest fruits as soon as it ripens.
- ☆ Remove the infected plant debris.
- ☆ Maintain well drained field.
- ☆ Crop rotation with non solanaceous crops for three years.
- ☆ Avoid wounds to the fruits during harvest.

Chapter

27

Postharvest Damage to Fruits and Vegetables by Virus

T.K.S. Latha

Department of Plant Pathology, Centre for Plant Protection Studies, Tamil Nadu Agricultural University, Coimbatore – 3, Tamil Nadu

Fruits and vegetables are prone to many virus diseases under field condition. The extended damage caused by the virus diseases result in persisting symptoms on the harvested produces *viz.*, fruit pitting, malformation of the shape and size, discoloration of the skin *etc.* These types of symptoms will greatly reduce the economic value of the fruits and vegetables under market as that of postharvest damages caused by fungal and bacterial diseases.

Alfalfa Mosaic Virus

- Alfalfa mosaic virus (AMV) is a virus usually found on alfalfa where it does not cause many problems.
- The vector aphids transmit AMV from alfalfa to nearby vegetables like bell pepper and tomato. The infected bell pepper and tomato leaves show typical mosaic symptoms.
- One of the most striking symptoms is the necrotic rings and spots on the fruit. Some fruit may develop a solid brown necrosis over the surface.
- The fruits of tomato will result in more number of small pitting of the skin whereas, the bell pepper tastes bitter due to the virus damage.

Courtesy: **T.A. Zitter**

Tomato Spotted Wilt Virus

- Tomato spotted wilt virus (TSWV) is an important pathogens of tomato, pepper, tobacco and peanut worldwide.
- Thrips have to acquire the virus as larvae to be able to transmit it as adults. Once larvae are infected, thrips carry and transmit the virus throughout their entire lifespan.
- The symptoms are necrotic spots on leaves, stunting of plants, necrotic rings on immature fruit and chlorotic ringspot on mature fruit in tomato and capsicum.

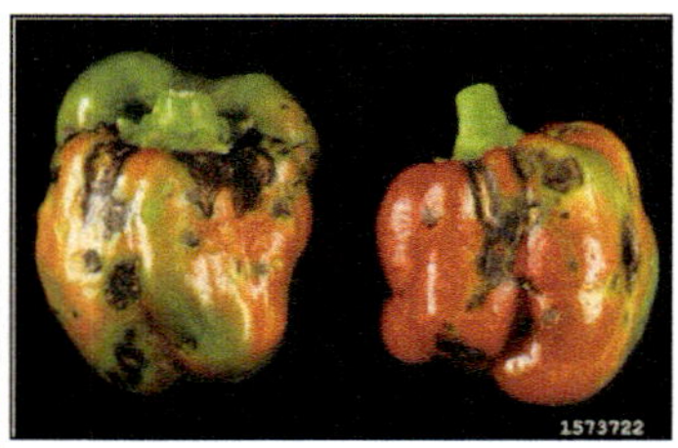

Courtesy: **J.L. Sherwood** *Courtesy*: **Gerald Holmes**

Tobacco Mosaic Virus

- Tobacco mosaic virus (TMV) causes damage on tomato crops also.
- They are not spread by insects but they are seed borne and spread by humans on hands, pruning tools, clothes.
- The infection results in fruit discoloration and malformation in the shape.

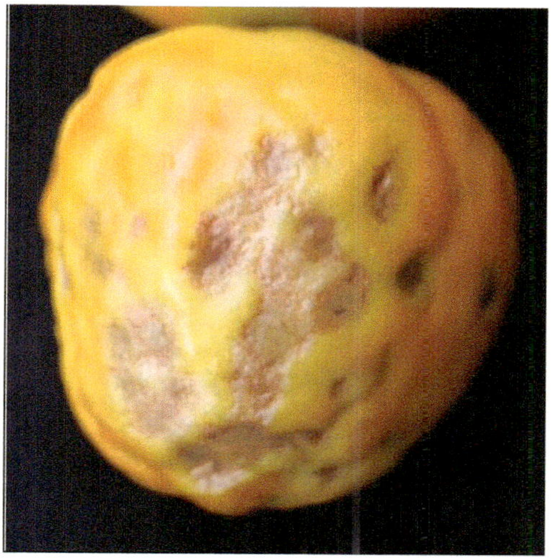

***Courtesy*: http://gardenofeaden.blogspot.com/2009/08/how-to-control-mosaic-virus-on-tomatoes.html**

Cucumber Mosaic Virus

- Cucumber mosaic virus (CMV) infects 1200 species in over 100 plant families and can cause significant economic losses in many vegetable and horticultural crops.
- Symptoms of cucumber plants infected early in the season are severely stunted and leaves are malformed, and fruit are unmarketable because of pronounced rugosity (roughness) on the fruit surface.
- Pumpkin is another cucurbit, that when infected at any early stage, will express severe foliar mosaic and the fruit will show a mosaic pattern and would be unmarketable.

***Courtesy*: T.A. Zitter**

Watermelon Mosaic Virus

- ✰ Watermelon mosaic virus (WMV) is aphid transmitted, and infection is limited to cucurbits. This virus is capable of infecting all commercial cucurbit crops.
- ✰ The foliage of affected plants shows strong mosaic, distortion, and deep leaf serration.
- ✰ Fruits are also malformed with knobby overgrowth.

***Courtesy*: https://www.picuki.com/tag/planpathologycy**

Papaya Ring Spot Virus

- ✰ Papaya ring spot virus (PRSV) infects papaya and cucurbits systemically.
- ✰ Symptoms on papaya are somewhat similar to those on cucurbits.

***Courtesy*: D. Gonsalves**

- In papaya, leaves develop prominent mosaic and chlorosis on the leaf lamina, and water soaked oily streaks on the petioles and upper part of the trunk.
- Severe symptoms often include a distortion of young leaves which also result in older plants that become infected do produce fruit that often show distinct changes in color and are deformed.

Pepper Mild Mottle Virus

- Pepper mild mottle virus (PMMoV) is a plant pathogenic virus that occurs worldwide on species of field grown bell, hot and ornamental pepper species that can be easily transmitted to healthy plants via mechanical inoculation or contact between plants or another medium that can carry the pathogen such as hands, gloves, and clothing.
- Typical symptoms include the chlorosis of leaves, stunting, and distorted and lumpy fruiting structures.
- Symptoms usually include various degree of mottling, chlorosis, curling, dwarfing, and distortion of the fruit, leaves, and even whole plants.
- The symptoms on fruit include: a reduction in size, mottling and color changes, and an obvious distorted and lumpy appearance.
- Also, many times brown necrotic streaks or splotches can be seen on the leaves and fruit.

***Courtesy*: https://www.cabi.org/isc/datasheet/43826**

Tomato Brown Rugose Fruit Virus

- Tomato brown rugose fruit virus (ToBRFV) infected tomato fruits can be discoloured, misshapen, turning yellow or brown with crinkling of the skin.

- The virus can readily spread to all plants in a crop. Due to the symptoms, the fruits of infected plants lose market value or are unmarketable.
- Fruit can develop chlorotic marbling which can appear and fruit may develop brown wrinkled (rugose) patches.

***Courtesy*: Aviv Dombrovsky** ***Courtesy*: Diana Godínez (MX)**

Zucchini Yellow Mosaic Virus (ZYMV)

- Foliar symptoms consist of a prominent yellow mosaic, necrosis, distortion, and stunting.
- Fruits remain small, greatly malformed, and green mottled.
- Color breaking of the fruit is another main symptom.
- On melon, dark green patches or blotches appear on the normally straw-colored surface of the fruit.

Knobbly and distorted pumpkin fruit;
Distorted and smaller zucchini fruit (left), healthy (right).

***Courtesy*: B. Coutts**

Chapter

28

Mushroom Diseases and Abiotic Disorders and their Management

G. Thiribhuvanamala, P. Latha, K. Angappan and A.S. Krishnamoorthy

Department of Plant Pathology, Centre for Plant Protection Studies, Tamil Nadu Agricultural University, Coimbatore – 3, Tamil Nadu

Mushroom cultivation one of the best hi-tech horticultural venture that utilizes the vertical space and is regarded as the highest protein producer per unit area and time, almost 100 times more than the conventional agriculture and animal husbandry. Mushroom being a fleshy vegetable crop is prone to attack by several insect pests, competitor moulds, fungal, bacterial and viral diseases. Mushrooms being an indoor crop, provide a suitable ideal environment for the multiplication of pest and disease causing organisms. Loss due to pest and disease incidence reaches sometimes 100 per cent in uncared farms. Dumping off spent mushroom compost invites harbouring of pests which is difficult to eradicate. Moreover, the growers do not practice proper hygienic conditions as they lack awareness on the life cycle of disease causing organisms. Most of the small mushroom farms are also not equipped with insect proof net and the spent mushroom beds are left dumped near the mushroom houses where insect pests multiply and gain entry in to the cropping houses. Inappropriate maintenance of temperature, relative humidity and carbon di oxide leads to abiotic disorders that pose both quantitative and qualitative yield losses. Chemicals are not used in managing pests and diseases due to residual problem of pesticides. Management of mushroom pests and diseases involve proper maintenance of sanitation in and around the cropping houses. This chapter deals with major diseases, competitor moulds and abiotic disorders that affect mushroom yield and their management.

Weeds, Moulds and Diseases

A) Weed Mold

i) Ink Cap

- *Coprinus comatus* in a common weed on mushroom beds due to poor quality straw, more processing of substrate during pasteurization, high moisture content of the substrate, more compaction of beds and poor ventilation.
- Ammonia production inside the cropping room favours development of inky caps.
- Initially rapid whitish mycelial growth is seen that produces long white thin stalk with whitish buds that turns black in colour with numerous black masses of spores that spread and cover the entire bed, thus arresting the growth of spawn, and development of young buds.

Management

- Avoid using damaged and old straw for bed preparation.
- Infected beds should be removed and destroyed immediately.
- Preparing beds with appropriate moisture content in substrate (60 per cent).
- Avoid chemical method of sterilization as this process lead to more weed growth.

B) Green Moulds

- Green moulds like *Trichoderma viride, Penicillium, Chaetomium* and *Aspergillus* is severe both in beds.
- The contamination leads to development of green color patches in the beds, which spreads quickly and entire bed is covered fully with green growth, which completely arrests the spawn running.
- This is due to improper sterilization of straw and bed preparation with more moisture.

Management

- Old straw to be avoided for bed preparation.
- Infected beds should be removed and destroyed immediately
- Preparing beds with appropriate moisture content in substrate (60 per cent)

Green moulds -*Trichoderma viride*

***Courtesy*: V.P. Sharma and Satish Kumar**

C) Fungal Diseases

- ✰ Few other pathogenic microorganisms infect the mushrooms and leads to quality and quantity loss

(i) Dry Bubble - *Verticillium malthousei* and *V. psalliotae*

- ✰ The disease is characterized by the deformation of the young mushrooms with distorted stipes and the formation of muddy brown and sunken spots on the cap of the mushroom.
- ✰ A grayish white mouldy growth may be seen on the pileus resulting in dryness and leathery appearance of the mushroom resembling a dry bubble.

Dry bubble infected mushroom

***Courtesy*: V.P. Sharma and Satish Kumar**

Management

- Maintain proper temperature, ventilation and humidity as this disease is favoured by temperature above 23°C, poor ventilation and high humidity.
- Perfect sanitary conditions should be maintained in growing houses and when the disease appears, the temperature should be lowered to below 28°C.

ii) Truffle - *Pseudobalsamia microspora*

- The disease is characterized by the appearance of cottony-weft of mycelium on the bed surface.
- These wefts soon become denser and develop into small, reddish brown and wrinkled stromatic bodies resembling a truffle.
- The infected beds usually have a peculiar disagreeable odour and the compost turns brown in colour.
- The yield of the mushroom crop is reduced significantly.

Management

- Casing soil used should be well sterilized.
- Excessive moisture in cased bed to be avoided
- Proper ventilation, temperaure and high humidity should be avoided in cropping room

iii) Cobweb - *Dactylium dendroides*

- A white silky growth of mycelium usually grows over the surface of the casing soil.
- It grows and covers the mushroom sporocarp and causes light brown depression on the cap.
- The older mycelium of the pathogen changes from silky to granular white and later becomes pink in colour.

***Courtesy*: V.P. Sharma and Satish Kumar**

- The infected mushrooms become discoloured and granular white.
- The infected mushrooms become discoloured and soft.
- Later on, such diseased mushrooms are engulfed in a cottony ball of mycelium.
- The disease is serious where year round cropping is practiced.
- Disease is more common in button mushroom.

Management

- Proper sterilization of casing soil is significantly important.
- As soon as the disease spots are located, complete casing oil should be removed and destroyed.
- Proper monitoring to prevent the spread of the disease.

D) Bacterial Blotch/Bacterial Pit/Brown Blotch

- This disease is caused by a bacterium, *Pseudomonas tolassi.*
- Infection leads to pale-yellow spots on the surface of the pileus, which later turn brown with distorted caps.
- Heavy watering at early bud stage leaves a film of water on the surface of buttons leading to browning and rotting, emitting a fowl smell.
- In addition - the tryoglyphid mites, beetles and water splash from the infected bed carry the pathogen from one bed to other.

Management

- Monitor the population of tryoglyphid mite
- Avoid spraying excess water to the beds.
- Removal and destruction of the infected beds periodically to avoid further spread.
- Spraying water mixed with bleaching powder @ 2 g/10 litres of water.

E) Viral Diseases

- Complex viruses cause a disease variously called the Brown disease/ watering stipe/X- disease/die back disease.
- Various symptoms are noticed on the infected mushrooms which may be open pilei, drumstick like fruit bodies and watery stipe.
- Sometimes, symptoms are not visible but may lead to severe reduction in yields of mushroom.
- Tissue isolated from such mushrooms show slow and depressed growth of the mycelium.
- In addition, the insect pests, phorid larvae and tarsenomid mites, spores from infected mushroom are aid in spread of the disease

Abiotic Disorders

- Unfavourable environmental conditions like in adequate or high level of substrate moisture, pH of the growing medium, bed substrate and relative humidity in the spawn running and cropping room, CO_2 accumulation in cropping rooms bring abiotic disorders in mushroom.

1. Stroma

- Appearance of discrete dull white or slight yellowish patches or aggregations of mycelia seen as a dense fluffy layer of tissue on the surface of the substrate or casing layer is called stroma that prevents free air exchange in the beds there by leading aggregation of mycelium that results in putrifying smell of the beds which invites flies.

2. Hard Cap and Open Veil

- Caps open prematurely and the gills become rudimentary or poorly developed.
- Toxic fumes, hot fumes may result in hard cap, sometimes become malformed where the cap size is smaller than the stem diameter.
- Open veil formation means premature opening with abnormal gill development which occurs due to water stress.

3. Hollow Stem and Brown Pith

- Water stress in the beds and less humidity in the growing room results in the formation of central hollow in the stipe.

4. Rose Comb

- Large swellings or humps appear on the mushroom cap initially and later gills appear over the cap or on any part of mushroom that looks like spongy.
- Such mushrooms become twisted, split and brown.

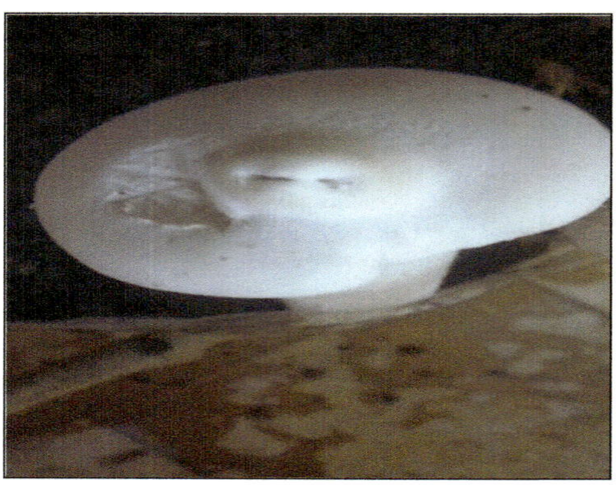

***Courtesy*: V.P. Sharma and Satish Kumar**

5. Scales or Crocodile Skin Formation

- Scaling is quite common on the pileus which may be easily peeled off.
- Mushrooms become shrunken with less weight.
- Low levels of relative humidity and increased temperature coupled with dry air in the cropping room increases scaling.
- This is seen in oyster, milky and button mushroom.

***Courtesy*: V.P. Sharma and Satish Kumar**

6. Long Stemmed Mushrooms

- Long stipeses with small cap appear due to excess CO_2 accumulation and poor light in the cropping room.
- Providing proper aeration can solve the problem.

***Courtesy*: V.P. Sharma and Satish Kumar**

7. Browning and Drying

- Browning and drying of the pin heads occur due to high temperature and less humidity. This is seen in oyster, milky and button mushroom.

Integrated Disease Management

Since mushrooms are used in salads, as vegetables by both children and aged people, use of pesticides should be avoided. However, strict sanitation must be, followed in spawn making and cropping houses to avoid much loss due to pest and disease attack. Pests also favour spread of many diseases. Hence, management of pest also should be taken care. The general hygiene and organic means of pest and disease management is discussed here.

- Selecting good quality substrates recently harvested paddy straw without insect and pest attack.
- Proper pasteurization of compost and casing material.
- Constructing mushroom sheds in elevated area and partially shaded area (trees); prevent water stagnation near sheds.
- Avoid using contaminated and over matured spawn for bed preparation.
- Prepare mushroom beds with moisture content around 60 percent; avoid more than 65 per cent moisture.
- Avoid chemical method of sterilization as this process leads to more weed mould growth and creates health hazards.
- Fixing insect proof nets in the windows of cropping rooms.
- Fixing insect traps to attract the flies or apply grease or oil on to yellow cards and hang it in sheds.
- Remove the infected beds periodically to avoid further spread of disease.
- Disposal of spent compost away from the cultivation yard.
- Avoid excess water to beds, always spray water to beds after harvest of mushrooms.
- Spray water mixed with bleaching powder @ 2g/10 litres of water on floor of mushroom house to manage bacterial growth.
- Apply neem oil (1 per cent) on the holes of polybags during bed preparation.
- Maintaining required relative humidity (70 per cent) to avoid scaling and drying of mushrooms.
- Maintaining ventilation to avoid long and thin stemmed mushrooms, small capped mushrooms.
- Dumping of spent waste near cropping room to be avoided.

Chapter

29

Cold Storage for the Management of Postharvest Diseases of Perishables

V. Sendhilvel and K. Prabakar

Department of Plant Pathology, Centre for Plant Protection Studies, Tamil Nadu Agricultural University, Coimbatore – 3, Tamil Nadu

Postharvest Loss

The term "Postharvest losses" is defined as quantitative and qualitative loss of materials which are used for food preparation and processedfood in various postharvest operations. "Food loss" is defined too as food available for human consumption, but not consumed. Limiting postharvest losses (PHL) is a priority area of cooperation between FAO and world development banks. The prearrest glut and postharvest loss is a common phenomena in the seasonal production agriculture and horticultural crops especially fruits, vegetables, spices and plantation crops and greens *etc.* It is very difficult to pin point the postharvest losses of perishable, an idea of the estimated loss ranged from 30-45 per cent has been widely accepted.

Global quantitative food losses and wastes during the year 2019 are around 30 per cent for cereals; 40–50 per cent for root crops and fruit and vegetables; 20 per cent for oilseeds, meat, and dairy products; and 35 per cent for fish. Every year, consumers in rich countries lose almost as much food (over 220 million tons) as the total net food production in sub-Saharan Africa (around 230 million tons). The entire amount of food lost or wasted annually is equivalent to more than half of the world's annual crop production (about 2.3 billion tons per year). (Barbara Sawicka, 2019).

India is the largest producer of fruits and second largest producer of vegetables in the world. In spite of that, per capita availability of fruits and vegetables is quite low because of postharvest losses which account for about 25 per cent to 30 per cent of production. Besides, quality of a sizable quantity of produce also deteriorates, by the time, it reaches the consumer. Most of the problems relating to the marketing of fruits and vegetables can be traced to their perishability. Perishability is responsible for high marketing costs, market gluts, price fluctuations and other similar problems. At low temperature, perishability is considerably reduced and the shelf life is increased and thus the importance of cold storage or refrigeration. The first cold store in India was reported to have been established in Calcutta in 1892.

Status of Cold Storage and its Potential in India

The estimated annual production of fruits and vegetables in the country is about 130 million tonnes. This accounts for 18 per cent of our agricultural output. Due to diverse agro climatic conditions and better availability of package of practices, the production is gradually rising. Although, there is a vast scope for increasing the production, the lack of cold storage and cold chain facilities are becoming major bottlenecks in tapping the potential. The cold storage facilities now available are mostly for a single commodity like potato, orange, apple, grapes, pomegranates, flowers, *etc.* which results in poor capacity utilization. A total of 136 cold storage units are functioning in Tamil Nadu.

Storage Conditions

For designing a cold storage, product storage conditions must be defined in terms of critical storage conditions of temperature, relative humidity, presence of CO_2, ethylene, air circulation, light *etc.* In absence of research data for Indian conditions, it is recommended to adopt commodity storage conditions as prescribed by Commodity Storage Manual of WFLO.

Temperature Range

The temperature in the multi commodity cold store chambers should be kept within + 1°C of the recommend temperature of the produce being stored. For storing at temperatures close to freezing point of the commodity, for increasing storage life, even a narrow range may be needed.

Humidity Range

The humidity (RH) is again dependent on the produce storage requirements and may vary from 95 to 98 per cent RH in case of fresh fruits and vegetables like grapes, kiwi fruit, carrots, cabbage *etc.* and lower in the range of 65-75 per cent RH in case of onion and garlic.

CO_2 Level

It should be not more than 4000 PPM during loading and 2000 PPM during holding.

Loading Rate

Generally the refrigeration system capacity is based on 4 per cent to 5 per cent loading rates of the total cold store capacity. The loading pattern is also a design consideration for sizing the storage chamber capacity for optimal utilization and performance

Pre-cooling Time

Minimum of 4-6 hours for pre-cooling to 7/8 th cooling time as recommended for majority of fresh fruits and vegetables. However in case of fresh produce like carrots, apples *etc.* meant for long/medium term storage, which are directly cooled and stored in the cold rooms, the cooling period, can be up-to 20 hrs per day and should meet the requirements specified in the commodity storage manuals.

Air Circulation

Multi Commodity stores should be design to provide an air flow of 170 CMH per metric ton of product, based on maximum amount of product that can be stored in each chamber. This is essential for rapid cooling of the produce. However the system should be designed to reduce air flow to 34 to 68 CMH per metric ton of product after the produce has reached the storage temperature.

Storage Conditions and its Groups for Storage of Fruits and Vegetables

Group No	*Name of the Fruits and Vegetables*	*Postharvest Pathogens Controlled*	*Tempe-rature*	*Relative Humidity*
I	**Products in this group Produce ethylene.** Apples, grapes (without sulphur dioxide) parsnips, apricots, horse-radish, peaches, pears, Barbados cherry, leeks persimmons, beets, plums, berries (except cranberries), loquat, pomegranates, cashew apple, lychee, prunes, cherries, mushrooms, nectarines radishes, figs (not with apples) oranges (Florida and Texas) rutabagas turnips	*Penicillum* spp., *Botrytis cinerea, Alternaria* spp, *Rhizopus* spp. (mostly *R. stolonifer*), *Penicillium* spp., *Cladosporium* spp.	0 to 2°C	90-95 per cent
II	Products in this group are sensitive to ethylene. Peas, asparagus, escarole pomegranate, bean sprouts, grapes (without sulfur dioxide), radishes, berries (except cranberries) kiwifruit rutabagas, leafy greens cabbage. Lettuce. spinach, carrots, Sweet corn, Cauliflower mushrooms turnips, celeriac onions, chestnut, mushrooms,	Various *Erwinia* spp., *Rhizopus* sp., *Penicillum* sp, *Bacillus* spp. *Pseudomonas* spp. and *Xanthomonas campestris*	0 to 2°C	95-100 per cent

Group No	*Name of the Fruits and Vegetables*	*Postharvest Pathogens Controlled*	*Tempe-rature*	*Relative Humidity*
III	**Moisture will damage these products** Garlic and onions	Various *Erwinia* spp,. *Aspergillus niger, Fusarium oxysporum* f. sp. *cepae, Colletotrichum circinans*	0 to 2°C	65-75 per cent
IV	Cactus leaves, lemons, cactus pears, oranges (Calif. and Arizona)	*Penicillium* spp., *Trichoderma* sp., *Botrytis cinera, Sclerotrnia* spp.	4.5°C	90-95 per cent
V	**Sensitive to chilling injury** Malanga squash, okra tamarind, cucumber olive taro root, eggplant peppers, potatoes, storage	*Erwinia* spp. *Fusarium* spp. *Phoma exigua* var *exigua, Rhizoctonia solani* (sclerotial state) *Helminthosporium solani*	10°C	85-90 per cent
VI	**Sensitive to chilling injury** Papayas, avocados, grapefruit passionfruit, guava, pineapple, bananas water melon, jackfruit potatoes, pumpkin, lemons, mangoes, sugar apple, tomatoes, ripe ginger root	*Colletotrichum* spp., *Phomacaricae, Rhizopus stolonifer, Colletrotrichum gloeosporioides, C. acutatum, Lasiodiplodia theobromae*	13 to 15°C	85-90 per cent
IV	Pears (for ripening) tomatoes, mature green white sapote	Various *Enurnia* spp., *Bacillus polimixa, Pseudomonas* spp. and *Xanthomona*, scampestris, *Botrytis cinerea, Fusarium* spp., *Sclerotium rolfisi*	18 to 21°C	85-90 per cent

Many preharvest factors directly and indirectly influence the development of postharvest disease, even in the case of infections initiated after harvest. A wide variety of fungal and bacterial pathogens cause postharvest disease in fruit and vegetables. Some of these, infect produce before harvest and then remain quiescent until conditions are more favourable for disease development after harvest. The cold storage is the feasible method for the management of postharvest disease in perishables to prevent loss.

References

Anon. 1997. Annual Report. Serdang, Selangor, Horticulture Research Centre, Malasian Agricultural Research and Development Institute(MARDI).

Bashi, E., Ben-Joseph, Y. and Roten, J. 1982. Inoculum potential of *Phytophthora infestans* and development of potato late blight epidemics. *Phytopathology*, 72**:** 1043-1047

Barbara Sawicka.2019. Postharvest Losses of Agricultural Produce. Eds. W. Leal Filho *et al.* (eds.), Zero Hunger.Springer Nature Switzerland AG. https://doi.org/10.1007/978-3-319-69626-3_40-1

Best, R.J. 1968. Tomato spotted wilt virus. *Advances in Virus Research,* 13*: 65-146.*

Brook P. J. (1968) Pre-harvest lenticel spot of apples, New Zealand *Journal of Agricultural Research*, 11:2, 237-244.

Bryan, M.K. 1930. Studies on bacterial canker of tomato. *Journal of Agricultural Research*, 41: 825.

Burr, TY.J. and Schroth, M.N. 1977. Occurrence of softrot *Erwinia spp.* in soil and plant material. *Phytopathology*, **67:** 1382-1387

Calo, R.P. 1957. Anthracnose of tomato. *Philippine Agriculture*, 41: 273.

Chand, R. 1996. Grapevine bacterial canker and its management. In: *Advances in Diseases of fruit Crops in India* (*ed.*) S.J. Singh, pp. 161-172, Kalyani Publishers, Ludhiana, India.

Commenil, P., Brunet, L. and Audran, J.C.1997. The development of the grape berry cuticle in relation to susceptibility to bunch rot disease. *Journal of Experimental Botany,* ***48:*** *1599-1607.*

Cooley, D., D. Handley, S. Schloemann, and W. Wilcox. 1998. Disease Management and Physiological Disorders. Chapter 9 in *Strawberry Production Guide for the Northeast, Midwest, and Eastern Canada.* Cooperative Extension, Ithaca, NY.

Coplin, D.L. 1980. *Erwinia carotovora var. carotovora* on bell peppers in Ohio. *Plant Disease Reporter,***64:** 191-194.

Dasgupta, M. K. and Mandal, N. C. 1989. *Postharvest Pathology of Perishables.* Oxford and IBH Publishing co. Pvt Ltd, New Delhi, Bombay, Calcutta, pp. 623.

Dang, K. T. H., Singh, Z., and Swinny, E. E. (2008). Impact of Postharvest Disease Control Methods and Cold Storage on Volatiles, Color Development and Fruit Quality in Ripe "Kensington Pride" Mangoes. Journal of Agricultural and Food Chemistry, 56(22), 10667–10674. doi:10.1021/jf801270a

Dennis, C. 1983. Soft fruits. In: *Postharvest Pathology of Fruits and Vegetables* (ed.) Dennis, pp. 23-42, Academic Press, London.

Dennis, C. 1983. Salad crops. In: *Postharvest Pathology of Fruits and Vegetables* (ed.) C. Dennis, pp.157-177, Academic Press, London.

Dennis, C. 1987. Fungi. In: *Postharvest physiology of Vegetables* (ed.) J. Weichmann, pp. 377-411, Marcell Dekker, Inc., New York.

Dennis, C., Browne, K.M. and Adamicki, F. 1979. Controlled atmosphere storage of tomatoes. *Acta Horticulturae,* **93:** 75-83.

Dennis, C. and Davis, R.P. 1980. Fungal spoilage of stored tomatoes. *Biennial Report of Field Research Institute,* 1977 and 1978, 44-45.

Eckert, J. W., and Brown, G. E. 1986. Postharvest citrus diseases and their control. p.315-360. In: *Fresh citrus fruits.* W. F. Wardowski, S. Nagy, and W. Grierson, (eds.). Avi Publishing, Westport, CT.

Eckert, J.W. and Sommer, N.F. 1967. Control of diseases of fruits and vegetables by postharvest treatment. *Annual Review of Phytopathology,* 5: 391-392.

El-helaly A.F., Ibrahim LA. and El-arosi H.M. 1954.*Alexandria Journal of Agricultural Research*11 (2), 9-28.

Freeman, S., Katan, T. and Shabi, E. 1998. Characterization of *Colletotrichum species* responsible for anthracnose diseases of various fruits. *Plant Disease,* 82**:** 596-605.

Hartill W.F.T. (1991) Postharvest diseases of avocado fruits in New Zealand, New Zealand *Journal of Crop and Horticultural Science,* 19:3, 297-304.

Henz, GP. 2017. Postharvest losses of perishables in Brazil: what do we know so far? HorticulturaBrasileira 35: 006-013. DOI - http://dx.doi.org/10.1590/S0102-053620170102.

Hildebrand, M., Dickler, E. and Geider, K. 2000. Occurrence of *Erwinia amylovora* on insects in a fire blight orchard. *Journal of Phytopathology*, 148: 251-256.

Indra, N. P. Aruna and V. Ponnuswami.2012. *Disease Management of Horticultural Crops under protected cultivation.* ISBN: 978-81-7132-713-3 Pointer publishers, Jaipur.

Jamaludin, Tandon, M.P. and Tandon, R.N. 1973. Studies on fruit rot of apple (*Malus sylvestris*) caused by *Gliocephalotrichum bulbilium* Ellis. *Proceedings of NationalAcademy of Science, India*, B,43: 154-158.

Jarvis, W.R. 1962. The infection of strawberry and raspberry fruits by *Botrytis cinerea* Fr.*Annals of Applied Biology*,50: 569-575.

Jordan, v.w.l. 1978. Epidemiology and control of fruit rot of *Botrytis cinerea* on strawberry. *Pflanzen schutz-Nachrisht,* Bayer, 31:1-10.

Jeger, M. J., Green, E. S., Tresh, J. M., Johnanson, A., Waller, J. M. and Brown, A. E. 1995. In: Banana and Plantains, pp. 317-81. (Ed.) Gowen, S. Chapman and Hall, London.

Joy P. P. and Sindhu G. 2012. Pineapple Research Station (Kerala Agricultural University).

Kader, A. (1982). Postharvest technology of horticultural crops. Publication 3311, University of California, Davis, California.

Kawamoto, S.O. and Lorbeer, J.W. 1974. Infection of onion leaves by *Pseudomonas cepacia*. *Phytopathology*,64: 1440-1445.

Lewis-Ivey, M.L., C. Nava-Diaz and S.A.Miller.2004. Identification and management of *Colletotrichum acutatum* on immature bell peppers. Plant Dis. 88:1198-1204.

Logan, C. 1983. Potatoes. In: *Postharvest Pathology of Fruits and Vegetables* (ed.) C.Dennis, pp. 179-217, Academic Press, London.

Ludivine L.,Marc, C, Luc de L.D.B and Haissam J.M. 2010. Crown rot of bananas:Preharvest factors involved in postharvest diseases development and integrated control methods. *Plant Disease,* 94(6), pp.648-658.

Maas, J.L., ed. 1984. *Compendium of Strawberry Diseases.* APS Press, St. Paul, MN.

Mandal, N. C. 1981. Postharvest diseases of fruits and vegetables in West Bengal. In: Postharvest Pathology of Perishables (Eds) Dasgupta and Mandal N. C. 1989. Oxford and IBH Publishing Co Pvt Ltd, New Delhi, Bombay, Calcutta, pp. 623.

Maude, R.B. 1983. Onions. In: *Postharvest Pathology of Fruits and Vegetables* (ed.) C.Dennis, pp.73-101, Academic Press, London.

McKeen, C.D. 1973. Occurrence, epidemiology and control of bacterial canker of tomato in southwestern Ontario. *Canadian Plant Disease Survey*, 53: 127-130.

Miller. M.E. and Dillion, R.C. Jr. 1979. Survey of bulb diseases on onions in South Texas. *Journal of Rio Grande Valley Horticulture Society*, 33: 25-28.

Pantastico, E. B. 1975. Postharvest Physiology, Handling and Utilization of Tropical and Sub-tropical Fruits and Vegetables. AVI. Publ Comp. Inc., West Port, Connecticut, USA.

Philip, S. 2002. Fruit Crops, pp. 211. Kalyani Publishers, Ludhiana, New Delhi, Noida, Kolkata, Cuttack.

Quazi, M. H. and Freebaim, H. T. 1970. *Botanical Gazette* 131: 5-14.

Rushing, J.W., Keinath, A.P. and Cook, W.P. 1999. Postharvest development and transmission of watermelon fruit blotch. *Horti.Technology*, 9: 217-219.

Sharma,V.P and Satish Kumar.2011. Competitor Moulds and Diseases in Mushroom Production 155 and Their Management In: *Mushrooms - cultivation Marketing and consumption Biology*(Eds.Manjith *et al.*,) pp.155-174

Snowdon, A. L. 1990. Postharvest Diseases and Disorders of Fruits and Vegetables (coloured atlas). Vol. I, pp. 302. Univ. Cambridge. Wolf Publishing Co. Ltd, London.

Snowdon, A. L. 1991 Postharvest Diseases and Disorders of Fruits and Vegetables (coloured atlas). Vol. I, pp. 415. Univ. Cambridge. Wolf Publishing Co. Ltd, London.

Srivastava, M.P. and Tandon, R.N. 1971. Postharvest diseases of banana in India. *Indian Phytopathology*, 24: 115-118.

Srivastava M.P and Tandon R. N. (1971) Postharvest Diseases of Papaya, PANS Pest Articles and News Summaries, 17:1, 51-54, DOI: 10.1080/09670877109413251

Stover, R. H. and Simmonds, N. W. 1987. Bananas, pp. 468. Third Edn, Longman Scientific and Technical Essex.

Suharban, M., Geetha, D. and Nair, H. K. 2000. Disease of guava fruits, College of Agriculture, Vellayani, Kerala (Personal communication).

Suharban, M., Geetha, D. and Nair, H. K. 2000. Disease of guava fruits, College of Agriculture, Vellayani, Kerala (Personal communication).

Swinburne T.R.and Brown A.E. 1983. Transactions of the british mycological society 80, 176-178.

Tandon, R. N. 1967. *Final Technical Report*. PL-480 Scheme. Allahabad University, Allahabad, Uttar Pradesh.

Thangavelu, P., Sundaraju, P. and Sathiamoorthi, S. 2004. Journal of Horticultural Science and Biotechnology, 79: 664–668.

Verhoeff, K. 1974. Latent infections by Fungi. *Annual reviewof Phytopathology*, 12: 99-110

Waller, J.M. 2002. Postharvest diseases. In:*Plant Pathologist's Pocketbook* (eds) J.M.Waller, J.M.Lenne and S.J.Waller, pp. 39-54. CAB International, U.K.

Wardlaw, C. W. 1972. Banana Diseases including Plantain and Abaca, pp. 877. Longman, London.

Wills, R.B.H., McGlasson, W.E., Graham, D., Lee, T.H. and Hall, *e.g.* (1989). Postharvest: an introduction to the physiology and handling of fruits and vegetables, 3rd edition. University of New South Wales Press, Sydney.

Wuest, P., and G. D. Bengtson. 1982. Penn State Handbook for commercial mushroom growers. The Pennsylvania State University, University Park.

Index

C

R

S